The Making of Intelligence

MAPS OF THE MIND

STEVEN ROSE, GENERAL EDITOR

MAPS OF THE MIND

Steven Rose, General Editor

—

Pain: The Science of Suffering
Patrick Wall

The Making of Intelligence

Ken Richardson

Columbia University Press

New York

Columbia University Press
Publishers Since 1893
New York   Chichester, West Sussex

Library of Congress Cataloging-in-Publication Data
Richardson, Ken.
    The making of intelligence / Ken Richardson.
        p. cm. — (Maps of the mind)
    Includes bibliographical references and index.
    ISBN 0-231-12004-4 (cloth : alk. paper)
        1. Intellect.   I. Title.   II. Series
BF431 .R53 2000
153.9—dc21
                                        00-022519

Casebound editions of Columbia University Press books are printed on permanent and
durable acid-free paper.
Printed in the United States of America
c 10 9 8 7 6 5 4 3 2 1

First published by Weidenfeld & Nicolson Ltd., London

Contents

What is intelligence? This is a big question for a little book. It is also one that has been tackled many times before, without clearly agreed success. So why try it again? Well, the importance of the concept, both socially and scientifically, persists undiminished. Indeed, it may, for various reasons, actually be increasing in importance. This is due not only to the increasing numbers of idea-makers but also to the existence of equally powerful media for their dissemination.

Intelligence seems to me to be just such a powerful idea, and just such a subject of increasing popular dissemination. There are probably several reasons for this. One reason is a scientific one, including the pace of expansion of psychology and the neurosciences over the past twenty years, and the use of concepts of intelligence as a focus for activities within them. For example, many books have appeared in recent years claiming to have discovered how intelligence arises in the tissues of the brain; how it has been simulated in computers; and how "genes for IQ" have been identified. Such claims need to be checked. But another reason for the increasing importance of intelligence is a social one. Government and business leaders say we need it more and more for competitive success; the pressure on schools to find and cultivate it has increased accordingly; the widening social inequalities are increasingly viewed as either a cause or consequence of individual differences in intelligence, depending on your point of view. All these seem to be good reasons for another look at the subject, especially within the context of a more general series on the mind.

This complexity of issues also makes intelligence a difficult subject for a small book, especially if any attempt at originality or the resolution of questions old and new is to be made. Nevertheless, this is what I have tried, and this is the result. It would not, of course, have been possible without the painstaking work of all the idea-makers I have had the pleasure of encountering over a quarter of a century of wrestling with the subject. Some have a presence in this book; I hope the rest will recognize, in the nature of the product, my sense of gratitude. I would also like to thank the many other colleagues and friends for criticism and encouragement on this and previous work on the subject, especially my partner Susan who, over a similar period of time, and with acute awareness of the social import of these issues, has given me her constant support. Finally, I would like to thank Steven Rose and Peter Tallack—and most recently, the editors at Columbia University Press—for their more direct hands-on and painstaking guidance in helping me turn a rough draft into a more readable product. If it still isn't readable, the fault is, of course, entirely my own.

The Making of
Intelligence

The Many Faces of Intelligence

There has probably been a concept of intelligence, and a word for it, since people first started to compare themselves with other animals and with one another. We know this at least since thinkers first began to theorize about the nature of the mind. To the philosopher Plato in Ancient Greece, the concept was simple: intelligence is that which distinguishes the different social classes, and is unevenly endowed by God. So fixed and innate to the individual was this endowment, he said, that general improvement could be obtained only by restricting reproduction to the "rulers." These are the so-called eugenic measures of which we have heard many an echo in modern times.

Plato's pupil Aristotle was more egalitarian, arguing that people (other than slaves, which he barely considered to be human) were much the same in their faculty of intellect, differences being due to teaching and example (and their possession of this faculty alone distinguished them from the animals, which, he said, were deprived of it). The Romans, too, debated these matters, as did the saints and schools of the Middle Ages, the philosopher/psychologists of the Renaissance—indeed the debate has not ceased, up to the present time.

Such a concept, and a word (or words) for it, has been found in every known society, including contemporary tribal societies, in various parts of the world. In everyday discourse today, "intelligent" must be one of the most commonly used terms for describing people. Indeed, people tend to use the term remarkably freely to describe others, and seem to be able to spot it extremely quickly. Interview panels think they can find it, or not, in their candidates in half an hour of searching questions. Teachers usually have no hesitation in describing their pupils as intelligent or not (often in the form of euphemisms like "bright" or "dull"). Parents often look for telltale signs of it in their own children—according to some reports, even in the first few days of life! They see the cultivation of their children's innate intelligence as the main function of the school, and often worry about whether it is being exercised or "stretched" enough for its full development. Popular pressure groups that track and select gifted children want some means of identifying it, as part of an individual's "potential," at the earliest possible age so that special treatments can be designed to help fulfill it. Other public bodies, such as the recent National Commission on Education (set up by the British Association), stress how, under increased international competition, the cultivation of intelligence has become central to economic success.

This pervasiveness of the idea of intelligence is such that most people take it for granted and don't question it too much. What is most peculiar, though, is that most of the time these people won't really be sure what they are talking about: they "know" what intelligence is, but they can't say what it is. Ask those same parents, teachers, politicians, employers, and so on what they mean by intelligence, and the chances are that they will not get far beyond shallow euphemisms such as "bright," "quick," "brainy," "clever," "quick to learn," "smart," and what have you.

This paradox—of "knowing," yet not knowing, what intelligence is—has been shown in a number of more systematic studies. One approach has been to put clear statements about intelligence to people (sometimes, literally, in the street) and ask them for their levels of agreement or disagreement. For example, Italian social psychologists Gabriel Mugny and Felice Carugati gave statements such as the following to a group of parents (the numbers in parentheses give their average ratings of agreement on a scale from 1 to 7):

a. Intelligence is gauged by the capacity for abstract thought (2.41);

b. The brain is the birthplace of intelligence (5.06);

c. The development of intelligence is the gradual learning of the rules of social life (3.62);

d. The development of intelligence proceeds according to a biological program fixed at birth (3.22);

e. Everyone is intelligent in their own way (5.00);

f. Some people are born with more intelligence, others less (4.97); and

g. The child develops intelligence by his or her own activity (4.82).

Because many of these questions contradict one another, the substantial levels of agreement with all of them suggests that either people disagree a lot, or they are individually mixed up about intelligence, or both. The parents in this group were from Geneva and Bologna (and there were more than forty questions in the original study). But similar uncertainty has arisen in many other studies in Britain and the United States. Mugny and Carugati also showed how such opinion wavers with time and place, even in the same person. They thus described intelligence as a "polysemous" concept—many meanings serving many purposes—which one moment is a general property of humans setting us apart from all other animals, and the next is the most fundamental property separating individual humans, and thus is the grounds for discrimination in education, jobs, promotion, and so on.

It has also been shown how much casual judgment of other people's intelligence is often based on superficial signs such as verbal articulation, general self-presentation, and even facial appearance and physical height. On such grounds, people quickly jump to conclusions about the intelligence of different groups, as well: women, blacks, Irish, Jews, Asians, and so on, depending on where and in which period you live. Everyone, however, will be convinced that intelligence is the first prerequisite for success—in school, in jobs, in social and family relations, in marriage, and in life.

No doubt these more or less casual views of ordinary people have been much influenced by views of academic psychologists who have, over the years, not been slow in publishing their own views in popular media. What these have demonstrated, however, is that it is not just ordinary people who embrace the concept of intelligence so assuredly, while simultaneously exhibiting confusion about what it is. This same problem is found among psychologists, as well. For example, in one survey in the United States, Yale psychologist Robert Sternberg and his colleagues asked a large number of psychologists to state what they thought intelligence to be. There turned out to be little overlap in the substance of their responses. Of the twenty-five attributes mentioned, only three were mentioned by 25 percent or more of respondents. More than a third of the attributes collected were mentioned by less than 10 percent of respondents. If we're asking experts to describe edible field mushrooms so we could distinguish them from the poisonous kind, and the experts responded like this, we might consider it prudent to avoid mushrooms altogether.

Despite this confusion—and although some individuals look askance at it—the subject of intelligence in scientific circles is not confined to a few psychologists. Increasingly in the twentieth century it has also figured among the conceptual furniture of brain scientists, computer scientists, evolutionists, animal behaviorists, and anthropologists. And, among scientists as well as among people in general, it seems to be a remarkably flexible, and "handy" concept: a concept of convenience for filling in an argument at an intuitive level, but an embarrassment when brought out into the daylight. Intelligence often appears, indeed, to have the properties of the kinds of spirit agents to which people appealed in former times for determining all complex processes for which no other obvious mechanism could be found. For example, a "spirit of nature" pervading all of the physical world was once considered to determine the movements of Earth, the planets, the seasons, the weather, and so on. Divine judgment was once thought to intercede in battles to the death over social disputes. And a "vital principle" determining the difference between the living and nonliving created debate in biology for centuries. Such ideas were very common before they were exorcised and replaced by more tractable, material processes.

This already, I hope, suggests the scale of the task facing anyone who wants to answer, at all comprehensively, the question I am asking:

What is intelligence? To pin down spirits indeed is partly the feeling I get in facing up to the question in this book. Before reaching the end, though, I will try to be more positive and constructive than that. To put my approach in perspective, let me show you, in the rest of this chapter, the many domains this spirit has haunted and the many faces of intelligence with which we are presented. This will serve the purpose of hinting at the many matters I will be taking up more fully in the rest of the book. I hope this will provide an overview of the ground we will cover, as well as help you relate individual questions and issues to the field as a whole. I hope, too, it will persuade you why you should read further!

IQ and Intelligence

When most people think of intelligence they will probably think of IQ (intelligence quotient), which is someone's score on an intelligence test. For most of the twentieth century, a large branch of psychologists who call themselves psychometrists (or psychometricians) have claimed to be able to accurately estimate people's intelligence (from just after birth to any age thereafter) in a half-hour test consisting of short questions and simple problems. They claim not only to be accurately describing a natural pattern of variation in intelligence but also its origins. They say that IQ measures a fundamental "power" which they call "g": a kind of ghost in the mental machine. They aren't very sure what this power is, but they say that it varies in strength from person to person because of different genes (of which more below). As such, IQ scores are said to speak volumes about people's long-term potential, crucially informing the clinician, the teacher, the employer, and the politician.

Whether IQ tests can really do this is a debate that has stalked the whole of the twentieth century. And it still does. In the most recent controversy, the U.S. psychologists Richard Herrnstein and Charles Murray (of whom much more later) have again issued dire warnings about society's failure to ignore the results of IQ testing. To emphasize this, they present the stereotypes of groups at either end of the IQ ladder. In one group, individuals effortlessly float to the top of the pile, go to the best colleges, enter rewarding and prestigious careers, have six-figure incomes, constantly expand their options and freedoms, and generally have fulfilling lives. At the other end of the ladder, life goes steadily downhill; people collect at the bottom of society; poverty, drugs, and crime are rampant; these individuals become a waste and a menace, "creating fear and resentment in the rest of society." All of this inevitably, say Herrnstein and Murray, stems from levels of intelligence, as revealed by the IQ test.

Can all this be true? We will look at that question in chapter 2. But we should be acutely aware of its importance. The IQ model of ability has seeped into the public consciousness, largely encouraged, no doubt, by popular books with titles like *Test Your Own IQ*, the setting up of high-IQ societies (e.g., MENSA), and its use as a fundamental index of potential by educators, psychologists, and politicians. It is not a harmless, abstract matter. People have been denied immigration to countries, been sterilized, had their educational, social, and employment opportunities blocked, been denied marriage and reproduction, all on the basis of an IQ score. More recently, the IQ saga has taken on an air of science fiction. There have even been suggestions that people who score low on an IQ test should be given a cash inducement to agree to sterilization! Sperm banks have been established, each sample duly accredited with the donor's IQ. There has been talk of the possibility of cloning particular humans who were graced with high IQ, from samples of their DNA, with the expectation that exact copies will, as it were, spring into life—bodies, personalities, IQ, the lot.

The Genetics and Environment of IQ

Most people will be aware of "nature-nurture" debates surrounding intelligence. Genes (often popularly described as the biochemical codes for all our characteristics) enter into debates about human intelligence in two main ways. First, it has been claimed that genes constitute the basic formative agents of intelligence, determining the ways we think and the structure of our knowledge, for example. Second, it has been said that individual differences in intelligence are due to differences in genes. Because some "races"* (notably Blacks in the United States) score less than Whites, perhaps these "racial" differences are also genetic (or at least partly so)? All of this has caused major controversy throughout the twentieth century.

The argument was first articulated by the English biometrist Sir Francis Galton in 1869. It was repeated (essentially unchanged) by American psychologists Richard Herrnstein and Charles Murray, the authors of *The Bell Curve*, in 1994. Throughout the century, this picture has been repeated over and over again. It flourished, against one target group or another, throughout the Western world, before World War II. As the British neurobiologist Steven Rose and his colleagues explained, "the same argument was the basis for the German racial and eugenic laws that began with the sterilization of the mentally and morally undesirable and ended in Auschwitz." Sometimes the story has become quite grotesque. Over the past decade or so, the Canadian psychologist J. Philippe Rushton has argued that the Black deficit in IQ compared with Whites is part of a more general complex of traits including smaller brain size, bigger genitals, precocious sexual maturation, higher frequency of sexual intercourse, higher levels of sexual hormones, lower marital stability, higher infant mortality, lower altruism, higher criminality, lower mental health, and lower parental investment in children. In sum, this is a complex that puts Blacks at the back of a biological evolutionary trend in which Whites and "Mongoloids" (for example, Asians) are way out in front with all the virtues. Needless to say, a majority of scientists have dismissed this kind of thesis as ridiculous, although a number of psychologists have openly applauded it.

We will take a closer look at these arguments in chapters 2 and 3, which will also include analyses of the data said to offer evidence in support of them. These data have resulted, in the main, from studies of twins and adopted children. More recently, however, there have

---

*I put this term in quotation marks because, as I hope will become clear in subsequent chapters, there are no such things as human races in the way that biologists have usually thought of them; rather they are far more casual, social designations.

been rather sensational attempts to find "genes for IQ" in DNA, the very chemistry of the genes themselves. Although this search often grips the media, I shall attempt to show that the concepts and methods used entail some highly improbable assumptions about the nature of genes and environments relevant to the development of IQ. Of course, most people will presuppose that there is nothing essentially wrong with such a search: there is "bound" to be genetic variation underlying individual differences in intelligence (however we describe it). What I will hope to show in chapter 3, by looking at the role of genes in development as well as what happens to them in the course of evolution, is that the matter is far from being so simple.

Animal Intelligence

As we shall see, attempts to search biology for intelligence—or, conversely, the use of intelligence as a handy concept—has extended far beyond the genes. Most people see intelligence in other animals. In some studies, people have been asked to rate different common species for degrees of intelligence, and responses are readily forthcoming. Apes came first, then dogs, with cats close behind, followed by horses, cows, sheep, and so on, in that kind of order. This view has been widely shared by scientists. Indeed, Charles Darwin, in his *Origin of Species* (1859), described the necessity of something like intelligence as a faculty for guiding basic instincts, even among earthworms and beetles: "A little dose . . . of judgement and reason, often comes into play, even in animals very low in the scale of nature," he said.

In his *Descent of Man* (1871), Darwin also portrayed the common evolutionary roots that humans share with other animals, and suggested that "the mental faculties of man and the lower animals do not differ in kind, though immensely in degree." This automatically suggested that we could establish natural ranks of intelligence. Accordingly, British psychologist George Romanes published in 1892 a ladder—a "phylogenetic scale of intelligence"—covering groups of organisms from protoplasmic single-celled organisms to humans. This effort was based almost entirely on anecdotal evidence. But a coherent discipline known as comparative psychology soon sprang up, focusing on the relative performances of different groups of animals on simple laboratory tests, such as running mazes for food or escaping from cages.

The new evolutionary view thus presented intelligence as a common principle running through all animal life but gradually becoming amplified in humans. It also suggested a simple strategy for elucidating human intelligence. We can examine the nature of intelligence in its simpler—and hopefully more comprehensible—form in other animals, and thus describe it, and how it has increased in humans, more clearly.

This has been a most active line of research. During his incarceration on Tenerife during World War I, the German cognitive psychologist Wolfgang Kohler conducted some famous experiments in which he administered humanlike problems to chimpanzees in order to assess their performance. For example, he put fruit out of direct reach outside the cage, or suspended above the head, and observed the animals using sticks or stacking boxes to retrieve them. Kohler suggested that success seemed to depend upon a mental "reconstruction" of the problem situation by the animals, such as to lead to quite sudden insight into its solution. He suggested that this followed principles of gestalt psychology—a cognitive facility for constructing order and form out of sensory data—which he and his followers suggested was also a basic principle of human intelligence.

Such studies have continued to this day, and a wide range of tests of cognitive ability (some derived from human IQ tests!) have been devised. These have usually revealed a trend in improving performance from rats to cats to monkeys to chimpanzees and then to human children. Other studies have simply placed objects in animals' cages and looked for signs of intelligent action, such as exploratory manipulation or using an object as a tool to retrieve food. Such observations have, again, indicated a trend favoring great apes over monkeys over other mammals.

Have these studies helped us clarify the nature of intelligence? The gestalt ideas of holistic imaging and problem-solving are intriguing, but tell us little about the actual mental processes in any detail. Tables of rank and graphs of trends may be helpful up to a point, but they don't tell us what the differences really are. Some psychologists have argued that they are differences in learning ability, but this is just a switch of terminology without offering further light. Some have argued that, if evolutionary theory tells us that all animals are biologically adapted to their own peculiar niche in the world, then a single scale is quite inappropriate because we aren't comparing like with like. The idea of "learning within biological constraints" has become a widely shared point of view, applied equally to humans and other animals, and I will have more to say about this in chapters 3 to 5.

Finally, many investigators have attempted to reconstruct the evolution of human intelligence from the behaviors of our prehuman (hominid) forebears as construed from fossils, living sites, and habitats. Although, as South African paleontologist Philip Tobias has pointed out, this whole area is rich in speculation on the basis of (literally) fragmentary evidence, a rough but intriguing picture seems fairly well agreed. In the view of many, this picture starkly portrays the great quantitative and qualitative leap in intelligence that occurred in human evolution, striking a great gulf between us and even our closest biological relatives. I will be describing the nature and origins of this gulf in chapter 7.

Adaptationism and Evolutionary Psychology

The idea of intelligence as an "adaptation" has become particularly popular in recent years among psychologists looking for a biological underpinning to their inquiry. This idea is certainly not new, though. Darwin's contemporary Herbert Spencer had suggested in 1855 his *Law of Intelligence*: "the fundamental condition of vitality is that the internal state shall be continually adjusted to the external order." The brilliant work on genetics which followed in the twentieth century has also seemed to provide a plausible mechanism by which adaptations are obtained and encoded in genes. Because of this, adaptation has long been a buzzword in psychology.

Intelligence is "the faculty of adapting oneself to circumstances," said the French psychologists Francis Binet and Henri Simon, the authors of the first IQ test. Accordingly, we are increasingly urged to look to biology, and what it tells us about adaptation, for true enlightenment about intelligence. For example, in his book *The Nature of Knowledge*, the British evolutionist Henry Plotkin says we must think of "the human capacity to gain and impart knowledge" as "itself an adaptation, or set of adaptations," and that "we will simply not understand human rationality and intelligence, or human communication and culture, until we understand how these seemingly unnatural attributes are deeply rooted in human biology."

Thus it has come about that construing the specific circumstances in which our ancestors evolved half a million years ago is now considered to be the best way of telling us about the nature of the intelligence we have now. Such construing does not, of course, happen without certain preconceptions about the nature of intelligence to guide what we should be looking for. Because intelligence is seen as a set of fixed properties of the mind, so inquiry has focused on stable and repetitive conditions to which (it is thought) they were initially adaptive. Our intelligence has come to be seen as a system of relatively fixed mental organs or modules, each adaptive to different problem domains experienced in the course of evolution, and all specified and handed on across generations in a genetic code. Fatalistic warnings are issued by evolutionary psychologists such as Leda Cosmides, John Tooby, and cognitive scientist Steven Pinker about the way that human adaptations to stable and recurrent situations long past explain the current designs of our cognitive specializations. According to Pinker, our intelligence, having evolved in the Stone Age, is a misfit in the modern world of advanced technologies and complex civilizations.

As a result of such "adaptationist" stories, it has virtually become a test of scientific virility for psychologists, in the past decade or so, to declare themselves to be evolutionary psychologists. Sweeping claims have been made in recent years about the genetically determined "computations" in our brains determining the form of our intelligence, and individual differences in it. Human infants, it is now said, come into the world preadapted for every perceptual, cognitive, and social demand. Individual and group differences—including, for example, differences between the sexes or between different ethnic groups—have been "explained" simply by painting a plausible adaptationist picture for each of them.

The problem with the adaptationist view is that humans also appear to have evolved in circumstances that were exceptionally changeable (that is, nonstable and nonrepetitive). Indeed, even primitive humans seemed to have done much to change the environment. Accordingly, other theorists have pointed out that fixed modules for intelligence would be a liability under such conditions: indeed, they could not have evolved in the first place. This means that cognitive functions must take more abstract forms, which cannot, even in principle, be encoded in a linear string of DNA. As the animal behaviorist Richard Byrne suggested in his book *The Thinking Ape: Evolutionary Origins of Intelligence*: "The use of the term intelligence should be restricted to that quality of flexibility that allows individuals to find their own solutions to problems; genetical adaptations, by contrast, are fixed and inflexible, however well tuned to special environments they are." Needless to say, not even a cursory causal chain between strands of DNA and the full-blown human intelligence is described in any of the adaptationist accounts of mental organs, fixed modules, computations, or whatever, that are currently so popular. Here, again, we find intelligence being treated as a generalized thing or spirit agent (that which forms adaptations) rather than the object of substantive characterization. We shall be looking at these matters in much more detail in chapters 3 to 7.

The Brain and Intelligence

Nearly everyone believes that intelligence somehow resides in the brain. In the minds of the general public, as in those of many psychologists, intelligence is virtually synonymous with "brain power." The brains of high-achieving humans have therefore been minutely scrutinized for clues as to the nature of the magic ingredient. In this issue, two strands seem most prominent. Some investigators have simply settled for comparisons of brain size, both across species and across human individuals. Others, including psychologists and neuroscientists, have attempted to establish a principled relationship between brain structure and intelligence.

The first approach has transpired in various quantitative indices, mostly based on the volume of the cerebral cortex—the most recently evolved and expanded part of the brain—in relation to body size or to that of the subcortex, or older parts of the brain. The cortex, which comprises 80 percent of the brain's volume in humans but much smaller proportions in other species, is naturally thought to be the seat of what is most distinctive about human intelligence. The quantitative indices demonstrate the stupendous advantage that humans have over other species in this regard. All of this seems to testify to an impressive new "space" for intelligence. However, just what this space is for has attracted little description beyond general metaphors: indeed, as we shall see in chapter 8, psychologists and neuroscientists remain remarkably vague about the relationship between brain size and intelligence.

Attempts to measure brain size in living humans, as well as individuals after death, and relate differences in size to differences in personal achievement, have been a popular pastime for more than a century. Although sometimes transpiring in inflammatory theses about racial and other group differences in intelligence, the whole enterprise is fraught with methodological difficulties, problems of interpretation, and inconsistencies of data. These will also be discussed in chapter 8.

The other approach to telling us what bigger brains are for has been to look for associations between specific parts or aspects of the brain and specific cognitive functions. We now know that the brain is made up of billions of nerve cells, or neurons, sending out numerous long fibers, or axons, which make contact with other neurons, near and far, through special connections called synapses. A particular neuron may be receiving tens of thousands of connections from other neurons at the same time. Can such interconnectivity tell us anything about the nature of intelligence? Does any of it differentiate into specialized circuits within which the processes of intelligence can be located? Can we truly say that intelligence is "in" the brain?

Certain kinds of evidence have been taken to suggest that we can answer affirmatively to all these questions. Accidental or surgical lesions often result in rather specific deficits in cognitive abilities. Recordings from tiny electrodes inserted into single nerve cells in monkeys have sometimes shown selective responses to highly specific objects, such as a hand or a face. It has been suggested that therein lies the function of object recognition and categorization. Patterns of electrical and metabolic activity during the performance of specific cognitive tasks have been examined by means of electroencephalograms (EEGs), and new scanning techniques, such as positron emission tomography (PET scans), have opened up many exciting possibilities. This new field, called cognitive neuroscience, has already made several suggestions about the localization of intellectual functions in the brain. Some investigators have even claimed that we can identify the parts of the brain that "explain" individual or group differences in intelligence. For example, it has been claimed that the brains of men and women are "wired" differently, and that this explains some differences in intelligence.

All of this has created an important debate about whether or not we are really identifying bits of the brain that do specific intellectual functions, and whether they really tell us anything new about intelligence. We will take a closer look at these matters in chapter 8.

Metaphors for Cognition and Intelligence

The puzzlement surrounding the question of what intelligence is, and likewise, what its relationship with evolution and the brain is, comes to a head in attempts to describe its cognitive form in humans. Although knowledge representation and reasoning are acknowledged to be the core processes of intelligence, what forms these might take has proven difficult to specify in any agreed manner. A common resort has been to construct mechanistic accounts of intelligence, using metaphors borrowed from currently prominent technologies. This is a strategy of argument with a long history. The invention of machines during the industrial revolution inspired the building of mechanical models of the human body, which were soon extended to the mind. By the nineteenth century, Darwin was launching his theory in the teeth of natural theology (a prominent school of theology at that time), which argued that, just as people have placed mechanisms in complex instruments such as clocks, music boxes, and Babbage's calculating engine (a Victorian computer), so God has implanted mechanisms that explain the complex behaviors of humans.

Such metaphors have been used to describe intelligence in more recent times, too. In his book *Evolution of the Brain and Intelligence*, the American neuroscientist Harry Jerison (following an older idea of Kenneth Craik's) used the metaphor of an internal "working model" of the world experienced. The amount of detail it is possible to capture in such a model explains what bigger brains are for, and is what distinguishes species' intelligence. When IQ testers are asked about the intelligence they claim to be measuring, they will often imply a quasi-mechanistic "speed" or "power" function.

The advent of the electronic computer, of course, provided new metaphors for intelligence, and contemporary "computational" theories of intelligence have become very popular. These envision intelligence as the sum total of a number of mini-computational routines, operating like computer programs, that have built up in the brain, perhaps in the form of distinct modules, in the course of evolution. In addition, considerable effort has been devoted to the design of computers that can think like humans, especially in relation to particular tasks such as medical diagnosis and manufacturing. This enterprise is known as artificial intelligence. We will be looking further at such views in chapter 4.

Other recent theorists have taken the brain itself—the rich network of connections, and thus the potential for knowledge storage and stimulus processing—as the working metaphor for intelligence. Accordingly, there have been prominent attempts recently to examine the nature of the cognitions underlying intelligence by modeling it in simpler networks set up in computers, an approach known, appropriately enough, as connectionism. In this view, because all the brain's decisions rest ultimately on the pathways through the myriad nerve cell connections, and because these are modified, or weighted, by experience, then it is to those connection weights that we should look for an understanding of intelligence. However, whether these simplified networks are really like the brain, and whether they can tell us about human intelligence, have been highly controversial, and we will be discussing them at some length in chapter 4.

Perhaps the most comprehensive account of the cognitive form of intelligence is that of the developmental psychologist Jean Piaget, who pointed to the interactive (or epigenetic) construction of the embryo before birth as a possible metaphor for cognitive development. He also used various formal systems of logic as metaphors for the mechanisms of intelligence that develop. Piaget, too, often spoke of intelligence as adaptation, but of a more dynamic form than that envisioned by the biological adaptationists mentioned above. The key function of intelligence, he said, is to enable anticipation of change, and thus constructive action to utilize or nullify it. "One of the essential functions of knowing is to bring about foresight," said Piaget. We will be looking at his theory at some length in chapter 6.

As interesting and complex as Piaget's theory is, many theorists have found it to be limited by its (alleged) stress on a socially isolated intelligence in which, as the American psychologist Jerome Bruner famously put it, "the lone child struggles single-handed to strike some equilibrium between assimilating the world to himself or himself to the world." The sociohistorical account proposed by the Russian psychologist Lev Vygotsky (written in the 1930s but only relatively recently discovered in the West) has thus frequently been put forward as a more viable alternative to Piaget's genetic epistemology. It suggests that the form of intelligence, and thus its variation, arises not from internal construction but from the nature of social relations, and the social and technical tools through which they are organized, in the child's particular culture. This rivalry has generated some heated debate in recent years, which we will consider at some length in chapter 6.

Culture, Social Life, and Intelligence

Many theorists have stressed how humans live and act in social conglomerates, using shared language, technologies, rules for living, and other cultural tools. They have thus argued that this culture will have a fundamental and pervasive influence on both the forms of intelligence in people, and individual and group differences in it. One form of inquiry in this area has been to examine changes in aspects of culture over evolutionary and historical periods. For example, it is clear that the evolution of humans between about two million years and one hundred thousand years ago involved spectacular increments in both the productivity and sophistication of stone tools. Cultural evolution over the past few thousand years has, of course, been even more spectacular.

Another approach has focused attention specifically on the intricacies of social interaction. A view that has become popular since the late 1970s is known as the "social function of intellect" hypothesis. This idea suggests that the need to maneuver between social cohesion of the group and individual needs required a new order of individual smartness or cleverness, and that this is what largely constitutes the intelligence of humans and other primates. Observations of the social abilities of other primates, such as chimpanzees, have increased significantly in recent years, and these are sometimes described as a bridge to human intelligence. But, as Byrne points out, "any such bridge can never be more than plausible hypothesis."

While some have argued that the culture of humans and other primates is similar in quality (although perhaps different in degrees), others have suggested that they are radically different. For example, Vygotsky, on the basis of his own comparative studies, concluded that human children develop psychological functions that are completely absent in apes. Many contemporary psychologists have argued that culture is constitutional to intelligence, determining its forms, and their variations, and not just interestingly variable expressions of intelligence. For example, administering simple logical and other IQ-type tests to different cultures has shown not that they are deficient in intelligence, but that their intelligence can take forms widely different from our own. As Michael Cole of the Laboratory of Comparative Human Cognition in California concluded from a series of such studies, "the structure of thought depends upon the structure of the dominant types of activity in different cultures." I shall have much more to say about this in chapter 7.

Another issue has been about how different conceptions of intelligence are found in different human cultures. We now know, from observation and inquiry about how different groups use the word (or something like it) in everyday life, that they are often referring to something different from intelligence as perceived in Western soci-

eties. Finally, in this context, is the perennial issue about whether intelligence tests are culture-biased—in the sense that the permutations of
items selected, by virtue of their language or other familiarity of content, may unfairly favor one group of people over others. We will be
looking at all these issues in chapters 2 and 7.

Intelligence as Social Ideology

The perverting of scholarly work to support narrow social privilege is
not new in any field, but it has been, as might be expected, particularly
prominent in studies of aspects of the mind, such as intelligence.
Bruner recently pointed out how notions about what the mind is and
how it works have deep cultural and ideological roots, and most great
social changes and political advances have been accompanied by corresponding revolutions in conceptions of mind. Superficial stories
about animal intelligence, evolution, adaptations, brain power, and
simple measures of it, together with simple conceptions of genes and
environments, offer fertile resources for a whole range of ideological
concoctions where sharp conflicts of interest suggest them.

As I mentioned above, intelligence has probably been a vehicle of
social ideology and political apologia for as long as social classes have
existed, and the preservation of privilege has remained an ideological
imperative. In nineteenth-century Britain, the renewed stress on natural inequality (leading to the invention of the IQ test) coincided with
the establishment of new social alignments at home and imperialism
abroad. Spencer argued strongly that colonial exploitation was justified because "the minds of the inferior human races cannot respond to
relations of even moderate complexity." And the poor, having thus
proved themselves to be "unfit," should be denied all social welfare and
normal reproduction, and be allowed to die off.

The founders of the intelligence-testing movement in the United
States and Britain were mostly strong hereditarians and eugenicists,
who saw the IQ test as the key instrument in promoting their cause.
They, and many of their contemporary supporters, such as Rushton,
have been financially backed by The Pioneer Fund, which originated

in the 1920s as an organization for promoting research into "race" differences, and many of whose members have held explicitly racist, eugenicist, and pro-Nazi views.

One of the main arguments put forward in *The Bell Curve* is that a democratic belief in human equality is naive. They paint a picture of a genetic elite effortlessly ascending to positions of power and privilege in society, while, forever sinking down, is a festering genetic underclass. IQ testing should thus be used as a selection process for the allocation of privileged treatment in education and for jobs. Those who try to ameliorate this situation, they argue, fail to understand its causes: they are well-meaning but naive. Little wonder then that many psychologists agree with Sternberg that "the social policy recommendations of Herrnstein and Murray . . . do not follow from their data, but rather represent a separate ideological statement."

Other ideological aspects of studies and writings in intelligence have been pointed out. For example, one reason why the idea of adaptation has been so popular is that it suggests the kind of passivity of thought and action that is required of people to accept their social lot, in contrast to the active transformation of the world of which they are capable. Thus the metaphor of a structure/function "fit" between individual intelligence and the environmental conditions constraining it admirably suits workplaces in which people are often seen as mere productive units.

Other writers, such as Mugny and Carugati, argue that the whole concept of intelligence can be understood only as a social representation or, like gender (as opposed to sex), a social construct, the function of which is to rationalize a complex social world. They argue: "Intelligence, if such a thing exists, is the historical creation of a particular culture, analogous to the notion of childhood." To support this idea, they review evidence indicating that what counts as intelligence varies widely across cultures, changes over time in the same culture, is defined differently for children of different ages, and changes with the changing social experiences of individuals. Rather similarly, Sternberg says that "intelligence is invented . . . it is not any one thing. . . . Rather it is a complex mixture of ingredients. . . . The invention is a societal one."

Pinning Down Intelligence

We are beginning to see that the existing ground does not offer a firm foundation for anyone seeking to answer the question: "What is intelligence?" Indeed, it is a complex confusion. Most ordinary people seem to know what intelligence is, but it turns out that they aren't so sure. Most psychologists seem sure about it, but their conviction splinters into disparate fragments when they are asked to define it. IQ testers say they can measure it, but do they know what they are measuring? They say those measured differences reflect genetic differences at least as much as "environmental" differences, but how valid have their concepts and methods for demonstrating that actually been? Intelligence is said to be a general principle of animal life that was given a huge boost in the course of human evolution, but of what the difference consists, and why we have it, remains uncertain. This uncertainty is reflected in unanswered questions about what our huge brains are for. Evolutionary psychologists say they can tell us about the nature of intelligence from a view of the environment our ancestors faced (and that it consists of fixed modules), but how valid are their views of those environments (and thus its consequences)? Computational and connectionist psychologists claim to have identified the basic units and processes that do the work of intelligence, but are their models realistic, or just more simplistic metaphors for intelligence? Intelligence is, at least to some extent, a social and cultural characteristic, but what does this mean exactly?

It is clear that the concept of intelligence usually includes deep social and ideological assumptions (of the way that the social world should be, or is naturally). In this vein, we need to remember that we and our children are critically affected by the products of these debates. This is most starkly expressed in claims about group differences, especially the overtly racist theories that have broken the surface of scientific respectability many times this century. But they also cause distress of other sorts (perhaps sorts that we too readily take for granted). Many studies have shown how low IQ scores have caused children to be attributed with low potential by their teachers, and to suffer reduced aspirations and long-term damage to their self-esteem and self-confidence, and they easily drift into the kinds of social despair that are then attributed to their "intelligence genes." It is sad, indeed, that many of the products of debates about intelligence have produced a fatalism among the general public about their own and their children's potentials for personal and social fulfillment. It seems to me that these are matters that people really need to know about.

But despite this quagmire of issues, I will try to keep my approach straightforward. My task in the early chapters of the book will be to attempt to pin down the ghost of intelligence in its various guises, and also to attempt to banish it where I think that is necessary. If all that sounds negative, don't worry. I do think there are strong, positive messages to be drawn from a proper description of intelligence, especially its dazzling variety which we find in humans. That proper description—at least in outline—is what I try to advance in the second half of the book.

BIBLIOGRAPHY

This list provides more general citations and suggested readings appropriate to this chapter; more specific citations and readings can be found under relevant chapters.

Byrne, R. 1995. *The Thinking Ape: Evolutionary Origins of Intelligence.* New York: Oxford University Press. Considers the social bases of ape intelligence.

Cosmides, L. and J. Tooby. 1994. In L. A. Hirshfeld and S. A. Gelman, eds., *Mapping the Mind: Domain Specificity in Cognition and Culture.* New York: Cambridge University Press. Classic views on the evolutionary-modular nature of intelligence.

Jerison, H. J. 1991. *Brain Size and the Evolution of Mind.* New York: American Museum of Natural History. Latest edition of a classic in this field.

Johnson, M. H. 1997. *Developmental Cognitive Neuroscience: An Introduction.* Vol. 1 of *Fundamentals of Cognitive Neruoscience.* Norwell, Mass.: Blackwell. An introductory overview of this field.

Martin, L. M., K. Nelson, and E. Tobach. 1995. *Sociocultural Psychology: Theory and Practice of Doing and Knowing.* Learning in Doing series. New York: Cambridge University Press. Contains a number of excellent reviews of cross-cultural research and its implications for theories of intelligence.

Mugny, G. and F. Carugati. 1989. *Social Representation of Intelligence.* Tarrytown, N.Y.: Pergamon. Analyzes people's everyday conceptions of intelligence and their context-dependence.

Plotkin, H. C. 1997. *Darwin Machines and the Nature of Knowledge*. Cambridge: Harvard University Press. An evolutionary-adaptationist view of intelligence.

Robinson, D. N. 1995 (3d ed.). *An Intellectual History of Psychology*. Madison: University of Wisconsin Press. A well-known, highly readable account of the history of ideas in psychology, focusing on their underlying assumptions and political contexts.

Rushton, J. P. 1997. *Race, Evolution, and Behavior: A Life History Perspective*. New Brunswick, N.J.: Transaction. The only one of Rushton's works to cause storms of protest (see also *The Bell Curve* under Herrnstein and Murray in bibliography for chapter 2).

Tobias, P. V. 1996. "The Brain of the First Hominids." In J. P. Changeux and J. Chavaillon, eds., *The Origins of the Human Brain*. New York: Oxford University Press. A very readable article in a book covering many issues surrounding the relations between brain evolution and intelligence.

Vygotsky, L. S. and A. R. Luria. 1991. *Ape, Primitive Man, and Child: Essays in the History of Behavior*. Classic Soviet Psychology series. New York: Grove Press. Vygotsky and colleagues compare children's and apes' intelligence and development and lay down many of their basic theoretical principles.

IQ: The Misconstruction
of Intelligence

In 1996, in the aftermath of *The Bell Curve*—another inflammatory thesis about the IQ scores of different "races" (in chapter 3, I shall debunk the idea of "races" in the sense usually implied by the term)—the American Psychological Association set up a task force charged with summarizing "knowns and unknowns" in the intelligence debate. In its summary, the task force group points out that there are many different conceptualizations of intelligence, but by far the most influential is that expressed in IQ testing. Although I shall have occasion in several places to criticize the group's findings, there is little doubt that it is right about the dominance of the IQ test, and the conception of intelligence it embodies. This is why we need to deal with it in this chapter.

Of course, no matter how we describe intelligence, most serious investigators would testify to its astonishing variability. Its diversity far outstretches that pertaining to any other character in any species. Describing that diversity and its origins is one of the major challenges of cognitive psychology, and must be accounted for in any reasonable theory of intelligence. Given the social weight attached to the attribution of intelligence, though, how we describe it has always been a matter of some social conflict.

Such description is, of course, only one side of what must ultimately be a two-sided coin: first, to describe what it is, in the sense of what is common to the species; then, to describe the nature of its variation. Many have tackled the first task, the standard scientific challenge of painstaking observational studies, followed by construction of explicit theory and controlled theory testing. Jean Piaget (of whom more in chapter 6) is probably the outstanding example of that approach. The more controversial approach of the IQ testers, however, has been to bypass such demands. They have simply applied to intelligence the metaphor of physical strength or power: to these people, intelligence is something that can be described and measured as a single variable on which people can be ranked, as with physical strength.

This strategy is based on the presupposition that the variation of intelligence, like that of physical strength, is essentially determined biologically. There is nothing new about this idea. Plato, in Ancient Greece, said that the social ranks among citizens simply mirrored the degrees of intelligence implanted by God as part of their biological makeup. But it wasn't until the late nineteenth century that Sir Francis Galton introduced it as the basis of intelligence testing. Galton was Darwin's cousin, and a reading of *On the Origin of Species* in 1859 had convinced him about the importance of heredity in human affairs and the possible improvement of the race. Galton was convinced that differences in natural ability simply reflected differences in biological endowment. While overlooking the fact that biological inheritance and social inheritance (for example, of wealth, privilege, and social advantage) in families are closely intertwined, he repeatedly condemned "pretensions of natural equality." This belief led Galton, like Plato, to favor a eugenic breeding program for the improvement of society, and he wanted scientific measurement of natural ability, or intelligence, to further that end. Such measures would serve, he argued, "for the indications of superior strains or races, and in so favouring them that their progeny shall outnumber and gradually replace that of the old one."

The IQ test emerged from Galton's reduction of intelligence to a single biologically determined power and his attempts to perfect an instrument for measuring it. To this day, many people—including psychologists, as well as the general public—still view the test that ultimately emerged as a reasonable scientific instrument. Many psychologists have strongly promoted it as a measure not only of the current status of individuals but of their immutable, long-term potential. In the course of the twentieth century, this endorsement, in turn, has often influenced social policy, especially for the treatment of inferior "races" and social classes, and in employment and education. The test is still widely used, and the underlying conception of intelligence is still stoutly defended. Moreover, the idea of IQ has seeped into the public consciousness through alarmist warnings about the social cost of ignoring it, and through books with titles like *Test Your Own IQ*.

In this chapter, I want to do three things. First, I want to show what a peculiar kind of measure IQ is (this seems of the utmost importance because few people seem to be aware of it). Second, I want to show that claims about it being an index of innate intellectual potential, and thus a reliable, long-term predictor of ability and social worth, are manifestly false. Finally, I want to show that the methods and assumptions used in testing itself—and in the interpretation of test scores in areas like educational and personnel selection, assessments of mental deficiency and group differences—are scientifically misleading, and, at times, dangerous.

Galton's Measure

Galton was keen on measuring and ranking just about everything. Unabashed by the dangers of reducing complex, interacting phenomena to single factors, he would doubtless have found some means of ranking kaleidoscope pictures if he had thought it important to do so. His conviction that all aspects of ability could be described as values on single parameters was first expressed when he administered a questionnaire to members of the Royal Society in 1874. They were asked to grade their talents on a twelve-point scale for each of several items, and it is clear that not all found it as easy as Galton had anticipated. Charles Darwin was one of the respondents, and wrote in the margin of his questionnaire: "N.B.—I find it quite impossible to estimate my character by your degrees."

Inspired by Darwin's law of natural selection, Galton seemed to have made two immediate errors of logic. The first is that all important variation in traits simply must be biological, whereas Darwin had pointed out in *The Origin of Species* that the logic of natural selection suggests the opposite: that natural selection reduces, and ultimately eliminates, biological variation in characters important to survival. Variation only persists in characters of lesser importance, he said: "Generally, the characters which individually vary are of slight physiological importance."

Galton's second error of logic was his assumption that, by being biological, intelligence must be manifest in all individual activities, even quite simple sensorimotor tasks. Thus he reasoned that we could get an estimate of individuals' mental powers simply by measuring their sensory and physical ones. His research was given a tremendous boost when he set up an anthropological laboratory in London and was able to recruit large numbers of subjects who were administered tests of sensorimotor powers, such as reaction time, speed of hand motion between two points, strength of grip, judgments of length, and so on. As Galton expected, these tests revealed substantial individual differences. But how could he feel justified in calling them differences in natural ability, or the "indication of superior strains and races" he sought?

Here we encounter a problem that has dogged the intelligence-testing movement throughout its history. In Galton's ratings, as with any measuring instrument, we are asked to take a visible quantity as an index of a specific invisible quantity. How can we be sure that it really is? For all general and scientific measures, this is achieved either by direct perception (as in the measurement of length with a ruler) or a clear model or well-worked-out theory connecting the two quantities; for example, the color of liquid in a breathalyzer and the amount of alcohol in the bloodstream, or the height of a mercury in a sphygmomanometer and blood pressure.

The only model or theory that Galton had was his hunch about a biological power, with no clear scientific theory of how it worked. His solution to this hurdle was to become the hallmark of the "intelligence" test: it was to relate the visible measure, not with the underlying "biological" quantity at all, but with what he took to be another visible manifestation of the same power, namely people's existing social status and reputation. In effect, this means that we are able to take people's positions on the social scale as a "true" measure of intelligence, and the test score as a measure of "natural ability" insofar as it reflects such positions.

In effect, then, Galton's aim, and that of his followers, became simply an attempt to reproduce an existing set of ranks (social class) in another, the test scores, and pretend that the latter is a measure of something else. This is, and remains, the fundamental strategy of the intelligence-testing movement, and this gloss over the fundamentals of fully scientific measurement is what has dogged it throughout this century. Of course, the quantitative nature of the test helped to create an impression of scientific measurement. But the strategy shortcuts any theory about the actual entity to be measured, and this problem has loomed large in all the controversy surrounding the IQ test ever since.

As it turned out, however, differences in test scores were not associated with differences in social status in the way Galton had expected. Those who followed Galton, such as J. McKeen Cattell, an American who studied in Galton's laboratory, also found "disappointingly low" relationships between test scores and current social status. In other words, the tests didn't work as Galton and Cattel had hoped.

The significance of Galton's test, though, lies in the reasoning involved. The general strategy of using a collateral index (social status) as a surrogate for genuine measurement signals the origins of the intelligence-testing movement. The development of the strategy into a successful test simply awaited the invention of different kinds of test items and a different kind of collateral index.

Binet's Test

By the turn of the century, psychologist Francis Binet, working in the Paris area, had already spent more than a decade testing children of different ages on short, school-type test items. Like Galton, he thought that intelligence could be estimated by having subjects perform a series of quick tasks which, as it were, could serve as "samples" of an individual's intelligence. But he introduced two new aspects to his tests. First, the tasks he introduced were more "mental" than those of Galton. He devised these by first thinking of all the intellectual qualities that could be tapped by short questions and problems: general knowledge, memory, imagination, attention, comprehension of sentences and synonyms, aesthetic judgments, moral judgments, and so on. After devising vast numbers of short items thought to activate one or the other of these abilities, he administered them to children. Then he decided whether each item really measured intelligence or not, according to two other criteria: first, whether the average performance on an item increased with age; and second, whether the performance of children on an item matched teachers' judgments of the intelligence of the same children.

By selecting items in this way, Binet and his colleague Henri Simon produced their first "metrical scale of intelligence" in 1905. It contained thirty items, designed for children aged three to twelve years, arranged in order of difficulty. Here are examples of some of the items: naming objects in pictures; repeating spoken digits; defining common words; drawing designs from memory; telling how objects are alike (similarities); comparing two lines of unequal length; and defining abstract words (by describing the difference between such words as "boredom" and "weariness," "esteem" and "friendship").

Although similar in broad approach to that of Galton, it is important to stress how Binet's purpose was rather different. He wanted such tests to help screen out children in the Paris school system who might need special educational assistance, for whatever reason. His aim was a practical one; his method was pragmatic rather than scientific and was not based on any strong theory of intelligence. Indeed, he was later to rail against the views of Galton and his followers about fixed intelligence, accusing them of "brutal pessimism."

In using the Binet tests, the tester simply worked through the items with each child until the latter could do no more. Performance was then compared with the average for the age group to which the child belonged. If a child could pass the tests expected of a six-year-old, say, then the child was said to have a mental age of six. Binet used the difference between the mental age and the chronological age as an index of retardation (he considered two years to be a serious enough deficiency to require special attention).

In 1912, the German psychologist William Stern proposed using the ratio of mental age to chronological age to yield the now familiar intelligence quotient, or IQ:

$$IQ = \frac{\text{mental age}}{\text{chronological age}} \times 100.$$

Thus was born the first modern intelligence test. Within a few years, translations were appearing in many parts of the world. The circularity of the strategy of item selection and the very restricted context of its use were soon forgotten in the wake of the social uses for which the test was soon being deployed, especially in the United States.

IQ as a Weapon of Social Policy

Binet's strictures about the test were soon forgotten because other users soon realized that it actually "worked" in another sense: it predicted educational attainments—and with that, social class and "racial" differences—in the way that Galton's test hadn't. This property was, of course, predetermined by the very method by which the test was constructed: mix and match until the match was right. But another group of Anglo-American psychologists read more into it than that. These were, like Galton, staunch hereditarians, who were, especially during a period of massive immigration into the United States, worried about the biological future of the "race," and thus its intelligence. They seized upon Binet's method and turned it toward Galton's goal of a measure of "racial" fitness.

Chief among this group was Lewis Terman of Stanford University, who had developed a translation of Binet's test in 1916 as a means of identifying "mental defectives." He enthused over the way that using the test could help clear "high-grade defectives" off the streets, curtail "the production of feeblemindedness" and eliminate crime, pauperism, and industrial inefficiency. By using his IQ test, he said, we could "preserve our state for a class of people worthy to possess it."

Another leading figure was Henry H. Goddard, who had translated Binet's test into English in 1910. Goddard was concerned with feeblemindedness (as mental disability was then called in the United States) and fully subscribed to the idea that it reflected one tail of a continuum of genetically inherited factors. He argued that feebleminded people must not be allowed to reproduce, and found the IQ test to be a crucial instrument in the propaganda he raised.

Goddard was also concerned about the waves of immigrants then entering the United States, and persuaded the authorities to let him set up a testing station at the main immigration port, Ellis Island in New York City. He thus managed to ensure that all individuals were given the IQ test as they landed, using the tests in English through interpreters. In his account of the process, Goddard himself gives an ironic glimpse of the objectivity of the process. "After a person has had considerable experience in this work," he said, "he almost gets a sense of what a feebleminded person is so that he can tell one afar off." By these means, the country came to be told that 83 percent of Hungarians, 79 percent of Italians, and 87 percent of Russians were feebleminded. The amount of feeblemindedness he had exposed soon had psychologists pressing ardently for immigration controls, which eventually became law in 1924.

As many historians of science have pointed out, the subsequent growth of IQ was that of a blatantly racist tool. Those who followed Terman and Goddard in the IQ-testing movement advocated the sterilization of the feebleminded, a policy that was actually adopted by many states in America (see, for example, L. Zendeland's *Measuring Minds*), resulting in tens of thousands of surgical operations. Writing in popular journals and magazines, these psychologists came to present the IQ test as an evaluation of the actual genetic worth of people. It was a view that seeped into the psychology of the American public, and soon entered Britain, where eugenics ideas had sprung up and were being spread by psychologists.

Followers of Galton had, indeed, already organized an influential campaign in Britain. The Eugenics Education Society was founded in 1907 and its journal, *Eugenics Review*, was started a year later. The alleged degeneration of the "race" was soon being described in popular seminars, magazines, and the pages of newspapers, including the columns of *The Times* (London). While the psychologist Cyril Burt was arguing that intelligence was innate, and social class differences were due to heredity, Karl Pearson was insisting, in the *Encyclopaedia*

Britannica, that, "It is cruel to the individual, it serves no social purpose, to drag a man of only moderate intellectual power from the hand-working to the brain-working group." The new intelligence tests were just as energetically promoted as measures of "racial" and class stock, and in 1911 a report to the Board of Education recommended the use of them for the identification of defectives.

The campaign proved, in the long run, to be highly effective. With the advice of psychologists such as Burt, the report of the Consultative Commission on Education of 1938 (or, more popularly, the Spens Report) in Britain was able to declare:

> Intellectual development during childhood appears to progress as if it were governed by a single central factor, usually known as "general intelligence". . . . Our psychological witnesses assured us that it can be measured approximately by means of intelligence tests . . . it is possible at a very early age to predict with accuracy the ultimate level of a child's intellectual powers. . . . It is accordingly evident that different children . . . if justice is to be done to their varying capacities, require types of education varying in certain important respects.

Thus was founded the British 11+ examination and the selective school system which materially affected the lives of millions of children. Outside Britain, of course—especially in Nazi Germany—essentially the same assumptions were affecting lives in a more fundamental way.

Constructing the IQ Test We Want

Terman's test (the Stanford-Binet test, as he called it) subsequently became the standard for all tests, and is still (after several revisions) the most frequently used individual test in Britain and the United States. But it was (and is) constructed using basically the same method that Binet had devised. Terman introduced many more items (ninety in all), and a greater variety of items was used, including digit span, vocabulary, word definition, general knowledge, comprehension, and so on. But, as before, these are mostly minischool-type tasks ("Can

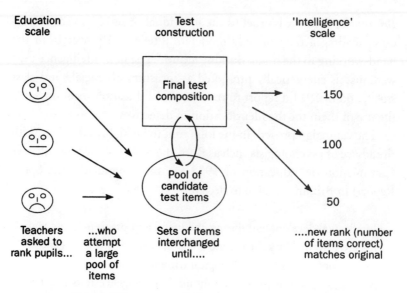

FIGURE 2.1. Steps in IQ-test construction.

you tell me, who was Genghis Khan?"; "What is the boiling point of water?"). And they were all selected, not on the basis of any theory that a particular item is tapping intelligence better than items not selected, but because those children who tend to pass them also tend to be those who are doing well in school.

In this regard, the construction of IQ tests is perhaps best thought of as a reformatting exercise: ranks in one format (teachers' estimates) are converted into ranks in another format (test scores; see figure 2.1). This is a common enough exercise in the present computer age. What is strange of course—and what even a computer-literate nine-year-old would think peculiar—is the claim of test constructors that the new format contains some "added" information not present in the first: they claim that the new format reflects ranks, not on the original format, but on the underlying, "biological" vein of intelligence.

It is through this strange logic that IQ testers have insisted that what they are providing is an independent estimate of "innate, general, cognitive ability" (the widely accepted definition offered by Burt). Generations of critics have pointed to the scientific illegitimacy of a claim that a correlation with one index automatically reflects a causal connection with another, but to no avail: the test furnished psychologists with a scientific-looking instrument in a discipline badly needing scientific respectability.

By this simple, but dubious, method, tests have proliferated enormously. In 1958, David Wechsler, chief psychologist at New York's Bellevue Psychiatric Hospital, introduced intelligence scales for adults and children that have come to rival the Stanford-Binet. The demand for labeling everyone for their innate intelligence, as in the mass testing of recruits to the armed forces and in educational and occupational selection, soon led to the design of "group tests." These consist of batteries of multiple-choice items that could be administered to large numbers of people in one sitting. Some of these tests were made up purely of nonverbal items, and so were considered to be measures of pure reasoning.

Perhaps the best known of these is Raven's matrices (see figure 2.2). This test was aimed to test "eductive ability," or what British psychologist Charles Spearman in the 1920s had viewed as the ability to grasp associations. However, the author of the test designed items on the basis of intuitive judgment—what seemed to him to be measuring eduction—rather than on any direct translation of theory. So what it really measures, or on what basis it really discriminates between people, is still a matter for much debate. As the most recent (1993) manual to the test admits: "there have been many fewer attempts to do as Spearman urged and examine the distinctive nature of eductive . . . abilities." Nevertheless, the claim continues to be made that the Raven's matrices (along with other nonverbal tests) is essentially a "culture-free" test, and that it is close to being a "pure" measure of the "power" Spearman had proposed. Both claims have been hotly disputed, as we shall see in chapter 7.

To this day, confusion about the theoretical bases of IQ persists. The technology of test construction, on the other hand, has improved by leaps and bounds. This reached the stage in the 1930s and 1940s that psychologists could predetermine patterns of scores on the basis of prior (more or less subjective) assumptions about intelligence, and build the test accordingly. This involves trying out large numbers of candidate items in advance and then selecting those that are going to produce the desired results.

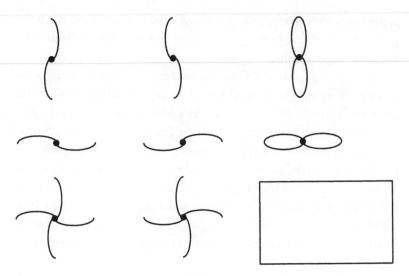

FIGURE 2.2. A typical matrix item, this one showing an "addition of two figures" rule. The correct entry for the blank space in the bottom right has to be selected from an array of six alternatives not shown here.

For example, it is assumed that IQ test scores, as measures of a supposed physical power, should be distributed in a population like height or physical strength, with many people having values at or near the average, and progressively fewer having values closer to the extremes. This is sometimes called the "normal" distribution (see figure 2.3). This is achieved in the IQ test by the simple device of including more items on which an average number of the trial group performed well, and relatively fewer on which either a substantial majority or a minority of subjects did well.

This distribution is, of course, the inspiration of Richard Herrnstein and Charles Murray's *The Bell Curve*, which I have mentioned a couple of times already. The authors seem to be convinced that, because it occurs with some measures, there is something natural and ubiquitous about it: "It makes sense," they say, "that most things will be arranged in bell-shaped curves." Indeed, IQ testers seem to have seen it as yet another manifestation of a supernatural force. Galton, who drew attention to it in 1889, spoke of "the wonderful form of cos-

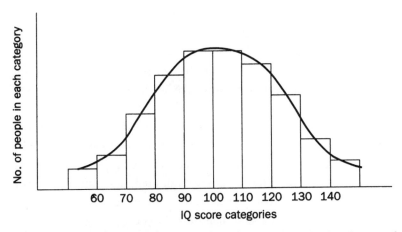

FIGURE 2.3. The famous "bell-shaped" curve obtained from the distribution of values of some physical characters, and built in to the IQ test.

mic order" expressed by it. So widespread is this view that nearly all the statistical analyses surrounding IQ testing (including the so-called genetic analyses, which I will discuss in chapter 3) have been designed to depend upon it. As the statistician R. C. Geary said in 1947, this is a deep-seated prejudice, based on the convenient fact that "when universal normality could be assumed, inferences of the widest practical usefulness could be drawn."

Even for achievement and psychometric scores, however, "universal normality" has little support in empirical studies. For example, in a report amusingly entitled "The Unicorn, the Normal Curve, and Other Improbable Creatures," Theodore Micceri examined the distributions of scores from large samples of testees on more than four hundred tests given in schools, universities, and workplaces across the United States. He found all to be significantly nonnormal. Others have shown that distributions of scores on tests of skill or proficiency, such as typing speed, are far from bell-shaped. Measures on almost all "basic" physiological processes, including visual acuity, resting heart rate, basal metabolic rate, and so on, are nonnormal in distribution. It needs to be stressed that, if the bell-shaped curve is the myth it seems to be—for IQ as for much else—then it is devastating for nearly all inference and discussion surrounding it. Yet the possibility is virtually completely ignored in the literature on IQ.

Another assumption adopted in the construction of tests for IQ is that, as a supposed physical measure like height, it will steadily "grow" with age, tailing off at around late puberty. This property was duly built into the tests by selecting items which a steady proportion of subjects in each age group passed. Of course, there are many reasons why intelligence, however we define it, may not develop like this. More embarrassing, though, has been the undesired, and unrealistic, side effect in which intelligence appeared to improve steadily up to the age of around eighteen years, and then started to decline. Again, this is all a matter of item selection, the effect being easily reversed by adding items on which older people perform better and reducing those on which younger people perform better.

In this way, too, specific patterns of group differences can be engineered "in" or "out" of test scores according to prior assumptions about what level of intelligence those groups are likely to have. For example, it was an accidental outcome of the first Stanford-Binet test that boys tended to score a few points higher than girls. At the time of the 1937 revision, it was debated whether or not this difference should be allowed to persist. It was concluded that the differences were due to "experience and training" rather than "native ability" and so it was duly eradicated by the appropriate selection of items.

This is, of course, a clear admission of the subjectivity of such assumptions. It is in this context that we need to assess claims about social class and "race" differences in IQ. These could be exaggerated, reduced, or eliminated in exactly the same way. That they are allowed to persist is a matter of prior assumption, not scientific fact. In all these ways, then, we find that the IQ-testing movement is not merely describing properties of people: rather, the IQ test has largely created them.

The "g" in the Machine

As mentioned above, and in chapter 1, IQ testing originated with little more than a hunch about what was being tested. Galton had viewed intelligence as an all-pervasive "natural ability" that was particularly prominent in the Victorian upper class. Binet, on the other hand, had viewed it as a kind of amalgam of abilities such as judgment, reasoning, good sense, and so on. As early as 1904, though, Spearman, another British psychometrist, was attempting to turn Galton's hunch into a respectable theory by examining the statistical interrelations among exam and test scores of various kinds. In this way, he came to report that children who did well or poorly on one kind of test also tended to do well or poorly on others, irrespective of the actual content of the tests. In other words, the scores were correlated.

Because we are going to talk about correlation a lot, let's be clear about what it means. A correlation is the tendency for scores on one measure to "co-vary" with scores on another; in other words, the measures vary together across cases. For instance, if we measure the heights and weights in a number of individuals (cases), we will find that those who vary to a certain degree from the average on height also tend to vary from the average to a similar degree in weight. This is called a positive correlation, because the covariation is in the same direction. A negative correlation can arise when the covariation is in the opposite direction. For example, there is probably a negative correlation between average hours of sunshine and millimeters of rainfall across counties of England. When Karl Pearson designed the first measure of correlation, around the turn of the century, he arranged it so the result was expressed as a proportion from zero to 1 for a positive correlation, or from 0 to -1 for a negative correlation. A correlation between height and weight in a typical sample of adults might be about 0.6, for example. This is the statistic Spearman worked with (and also modified and adapted in some ways).

Before going any further, though, we should note two crucial points. First, the correlation is not a measure of similarity between pairs of scores (that's simply not the way it's computed). Rather, it is a measure of the degree to which pairs of measures vary together, or co-vary, from their respective averages. Second, a correlation does not in itself imply a path of causation between the measures. All this is obvious when the pairs consist of things like sunshine and rainfall (for example, the lack of sunshine doesn't cause the increased rainfall), but the attribution of a causal connection becomes much more tempting with other measures, such as purported measures of intelligence and social status. A correlation between two factors may be due to an intermediate variable (the amount of cloud formation in the case of sunshine and rainfall), or it may be entirely accidental, with no causal connection whatsoever (for example, the population of Earth and the distance of Halley's comet from the earth over the past few years). Unfortunately, this oft-repeated caution frequently becomes forgotten.

Finally, it is as well to be aware that attempted replications of Spearman's work have not always produced correlations so high. In an article in the *British Journal of Psychology* in 1985, Raymond Fancher of York University, Ontario, reported how he reexamined Spearman's data and found them to be riddled with arithmetical mistakes, errors in calculations of correlations, and obscurities of method and reporting.

Table 2.1 shows some of Spearman's early results, taken from twenty-two boys in a preparatory school. Spearman reasoned that performances which correlate significantly like this are probably only attenuated expressions of a single underlying variable or factor. 'Accordingly, he leapt to the conclusion that "there really exists a . . . 'General Intelligence.'" He claimed that this was in the nature of an energy or power variable—the ability to draw inferences and make deductions—rather than the ability to learn, and he called it "g". The high correlations on tests across the board is evidence, he argued, that it is the energy source on which people vary, and such variation is determined by variation in genes. To the present day, psychologists have hailed this interpretation as the biggest single discovery in psychology. Indeed, most of the thesis of *The Bell Curve* rests on the existence or otherwise of g.

This reasoning—that correlated scores reflect a common ability—is just as common today. But it is just as important to point out that it is an act of the imagination rather than one of revelation. Even Herrnstein and Murray, the authors of *The Bell Curve*, seem to recognize this: "The evidence for a general factor was . . . circumstantial, based on statistical analysis rather than direct observation. Its reality therefore was, and remains, arguable." Thus they tacitly admit that their eight hundred or so pages of argument, dozens of graphs and tables, as well as their profound social-policy messages, are based on a

TABLE 2.1 Spearman's Reported Correlations Among Scholastic and
Sensory Measures

	1	2	3	4	5
1. Classics					
2. French	0.83				
3. English	0.78	0.67			
4. Math	0.70	0.67	0.64		
5. Pitch	0.66	0.65	0.54	0.45	
6. Music	0.63	0.57	0.51	0.51	0.40

gamble that their g is not a mirage. They are also tacitly admitting that the picture they attribute to g and its genes could be due to something else. But imaginative construction of intelligence on the basis of statistical analysis of test scores has gone much further than this.

The "Structure" of Intelligence

Spearman's reasoning was that variations in performance which are correlated are probably only attenuated expressions of a single underlying factor. Since the 1930s, Burt and his followers claim to have found more structure in intelligence than this. Now working on masses of IQ test scores, with more powerful computers, they claimed to find further intercorrelations between test scores that were not detected by Spearman. For example, those who tended to do well on one set of items, such as word definition, might tend to do equally well on sentence comprehension but less well on arithmetic items. This suggested a "group factor," which we might call a "verbal factor," as distinct from a "number factor." Thus a burgeoning "structure of intelligence" movement was started, and the debate generated has kept a lot of academic psychologists busy ever since.

The movement has only served to emphasize the subjectivity of IQ testing as such. Although this kind of factor analysis sounds like a straightforward and powerful technique, there is a snag: interpretation is largely intuitive. Correlations between sets of items can be made to appear or disappear, according to the prior assumptions used by an investigator in the computations. Prior assumptions about the nature of intelligence are required in deciding what kind of statistical analysis should be done, and so in deciding what is and is not a factor. In consequence, a wide variety of structures of intelligence have been found in the same sets of data.

For example, Burt decided that Spearman's general factor was underlain by two "group" abilities, which British psychologist Philip Vernon called verbal-educational (v-ed) and spatial-mechanical (k-m) factors (each in turn consisting of distinct but overlapping "special" abilities). This more hierarchical view (still with g at the top) became very popular in Britain and continued as an assumption in the construction of the most frequently used intelligence tests. Note that the factors are given names according to the content of the test items, not as a description of underlying cognitive mechanisms (which continue, of course, to remain shrouded in mystery).

Some American psychologists, however, tended to favor the Binet view of intelligence as a constellation of several different abilities. Thus American psychologist L. L. Thurstone "saw" a different pattern of correlations in IQ test data and, in 1938, arrived at results indicating seven separate factors in test scores, rather than a hierarchy of general and subfactors. These were described as: S, spatial ability; P, perceptual ability; N, numerical ability; V, verbal ability; M, memory; W, verbal fluency; and I, inductive reasoning.

A test was devised to emphasize, or "bring out," this pattern, again by selecting items with the appropriate scoring patterns in previous trials.

By using a different set of investigative spectacles, R. B. Cattell, who migrated from London to the United States in the 1930s, perceived a still different pattern of correlations in IQ test data. He explained how "in the mid-thirties some half dozen different lines of evidence converged in the present writer's thinking to suggest the disturbing idea that g might be two general factors instead of one!" Again using factor analysis, Cattell arrived at a distinction between a "fluid" intelligence and a "crystallized" intelligence. This distinction arises from the patterns of scores he claimed existed around the two different kinds of items in most tests: crystallized for acquired general knowledge and information (the vast majority of items in most tests), and "fluid" for reasoning itself (and best distinguished, he said, in nonverbal items, like the Raven's matrices, which were supposedly free from influences of learning and culture). Of course, how an energy (*g*) can be split into two like this isn't explained, but another kind of test emphasizing this pattern was duly constructed.

The burgeoning of separate factors continued in the model proposed in 1959 by J. P. Guilford, at the University of Southern California, which suggested up to forty factors. And so it has gone on. In the 1990s, John B. Carroll, an American psychologist who still believes in *g*, has suggested that more than seventy different abilities can be identified in IQ test scores. The disparate beliefs of psychologists in the different camps continue, as the following claims taken from volume 3 of *Advances in the Psychology of Human Intelligence*, and referring to essentially the same evidence, indicates:

> *H. J. Eysenck* [of London's Institute of Psychiatry]: "Psychometric studies have now pretty well resolved this dispute: there clearly is need for a general factor to account for the 'positive manifold' usually produced when IQ scores are intercorrelated."

> *J. Horn* [of the University of Southern California]: "There are good reasons for discounting the idea that there is a single, unitary capacity of general intelligence. Most of the evidence before us suggests that humans have several different intellectual capacities for which there is no functional unity."

We now have some idea of why even this subarea in the field of IQ testing is so contentious. As in the construction of tests themselves, you don't get what you see: you get what you want to see. What is perhaps most remarkable in all of this is that investigators are surprised that performances on such tests are correlated at all. The items are, after all, devised by test designers from a very narrow social class and culture, based on intuitions about intelligence and variation in it, and on a technology of item selection which builds in the required degree of convergence of performance. The fact that the supporters of the IQ test then claim that such patterns reflect real structures in nature ought to fool no one; alas, it fools them all the time.

What Do IQ Tests Measure?

In the absence of agreed definition or characterization, IQ testers have sometimes been asked to provide their own personal opinions of what is being measured. This was first attempted in a symposium in 1921 in which the editors of the *Journal of Educational Psychology* asked those prominent in the area of intelligence at that time to state what they considered "intelligence" to be. The diversity of answers received, and the absence of agreement among them, led to American psychologist E. G. Boring's half-joking, half-exasperated claim that "intelligence is what intelligence tests test."

This exercise was repeated by American psychologists Douglas Detterman, Robert Sternberg, and colleagues in the early 1980s. They wrote to a couple of dozen theorists, asking them the same questions that were put to the experts in 1921. This time, however, the investigators analyzed the results for frequencies of mentioned attributes. Of the twenty-five attributes proposed, only three were mentioned by 25 percent or more of respondents: half of the respondents mentioned "higher-level components"; 25 percent mentioned "executive processes"; and 29 percent mentioned "that which is valued by culture." More than a third of the attributes were mentioned by less than 10 percent of respondents in each case. So, it seems that experts' definitions of intelligence have changed over time, but there is still just as much disagreement among them.

Predictability of IQ Tests

Why do most psychologists continue to think that IQ tests are actually measuring something deeper than the imposed assumptions of test designers? Strangely enough, most of their conviction appears to rest on the fact that a child's IQ test score will predict school performance now or in the future. The group from the American Psychological Association, which we mentioned earlier, is obviously impressed by the fact that IQ tests predict school performance, pointing out that "the correlation between IQ scores and grades is about 0.50."

What seems to be remarkable is the disregard for how this kind of correlation comes about. As the American experts in educational testing R. L. Thorndike and E. P. Hagen explained in 1969: "The fact that intelligence tests correlate with academic achievement and school progress is unquestioned. From the very way in which the tests were assembled it could hardly be otherwise." If we devise IQ tests, as we do, to contain knowledge that has been learned in school, such as "In what continent is Egypt?", "Who wrote Hamlet?", or "What is the boiling point of water?", there should be no mystery about this correlation. Indeed, nor does it add anything to what we already know. It has long been demonstrated that teachers can predict with far greater accuracy, and in a fraction of the time, the future achievements of their pupils. Yet how much more seductive is the mystique of a scientific-looking test, and the mysticism of "general intelligence."

If the correlation between IQ and school achievement was due to an independent causal power, then we would expect it to be apparent throughout life, so that test scores should predict performance in all intellectually demanding situations, such as work. Of course, school attainment largely determines the level at which individuals enter the job market, so there is a built-in correlation between IQ and people's ultimate level of occupation. Again, this is a foregone conclusion.

But what about actual competence and performance in work? Here, the picture becomes very murky indeed. There have been few good studies of the association between IQ and job performance (the latter as estimated by things like peers' or supervisors' ratings, or samples of work). Such studies are, of course, hampered by uncertain suitability of tests, poor reliability of methods, and selectivity of reporting and citation.

Their interpretation is also often tortuous. For example, Herrnstein and Murray, the authors of *The Bell Curve*, say that the best predictors of occupational performance are broad ability tests. This could be because their own studies happened to be based on such a broad test. It involved a range of items described as arithmetic reasoning, word knowledge, language comprehension, mathematical knowledge, and numerical operations and was justified by the declaration that it is a "broad measure" of Spearman's g. On this basis, they report a correlation of around 0.5 between test score and job performance.

Ironically, though, Spearman was against broad tests (made up of an unscientific "hodge-podge" of items, as he put it) as a measure of g. Logic would surely also suggest that, if an IQ test is valuable as a measure of the elusive g (which is what Herrnstein and Murray argue), and it is really g that is important in job performance, then the narrower tests should be the best predictors. The Raven's matrices, specifically designed as a measure of Spearman's g, obtains much lower correlations with job performance (around 0.3). The authors of the 1993 test manual thus declare that the predictive validity of the test with regard to occupational success "is relatively low." In fact it is so low that the authors seem obliged to warn that "the popular notion of General Ability . . . does not merit the explanatory power and attention accorded to it by psychologists, managers, educators, and educational therapists"!

The group from the American Psychological Association, in its survey of the few studies available, put the raw predictive correlations between IQ and occupational success to be around 0.3 (itself, of course, a small association that would explain only a tiny proportion of variability in job performance between individuals). But they do not consider the possibility that even this association may be an indirect, secondary effect of schooling and socioeconomic status. For example, the boost to self-esteem, self-presentation, self-confidence, and so on might translate (initially at least) into better performance in the workplace. It has been shown that, when such background advantages are controlled statistically, the correlations between IQ and job success become negligible.

The idea that the (already small) correlation between IQ and job performance is due to such side effects is supported by the way that it declines over time. This is shown in several studies reported by anthropologist Allan Hanson of the University of Kansas in his book *Testing, Testing*. For example, scores based on extensive testing of servicemen in World War II found little association between those scores and their occupational success twelve years later. Other studies have shown that the correlation between IQ and job performance progressively reduces to zero among those who continue in the job for several years. Among the mediators of such secondary effects may be self-confidence, as the occupants throw off their doubts (acquired in school and elsewhere) about their own abilities. What does seem to be clear is that the idea of IQ as a measure of potential that manifests itself throughout life finds little support in properly controlled studies.

IQ and Cognitive Ability

Another way of questioning whether IQ is a test of basic cognitive ability important throughout life is to see whether people with high or low IQ perform better or worse in jobs requiring cognitively complex processing, or to compare performance on IQ tests with that on other cognitively complex tasks. Such means allow us to ask whether high IQ can be equated with complex cognitive ability. We shall see later, the answer to this question ought to be apparent from an examination of test items themselves, which depend largely on memory and general knowledge, which will obviously be culture-biased. Even the nonverbal items, like Raven's matrices, can be shown to be elementary in actual logical structure, and, if anything, are even more culture-biased than their verbal counterparts (see chapter 7).

But the question can also be answered in other ways. As we have just seen, after settling into a job a person's IQ is irrelevant to how well he or she is doing. A number of other studies have compared the IQs of highly creative or achieving groups with mediocre groups in the same occupation or profession, and found test scores to be irrelevant to group membership. American psychologists Anders Ericsson and Neil Charness have reviewed several studies, involving a wide range of occupations, showing that average IQ did not discriminate the more from the less competent in any of them. A more revealing approach in recent years, however, has been to analyze subjects' performances on the learning and/or control of complex dynamic systems, and then compare these with their IQs.

In a well-known study reported in 1986, American psychologists

Stephen Ceci and Jeffrey Liker explored the performances of regular bettors at a racetrack, who were asked to handicap horses in ten actual and fifty contrived races. Their analyses revealed two important things. First, they showed that, far from being guesswork or intuition, expert handicapping is actually a cognitively sophisticated enterprise in which the experts combine information from up to seven variables (including jockey weight and distance) at once. These combinations, moreover, often contained nonlinearities (for example, the increased effects of one variable, such as jockey weight, on another, like running speed, may not be the same for each increment of weight) and multiple interaction effects (very abstract relations in which, say, the association between jockey weight and running speed is itself influenced by the state of the ground, and this complex relation is further influenced by the race distance). Second, they found that performance on this cognitively complex task was unrelated to measured IQ. This led the authors to conclude that "whatever it is that an IQ test measures, it is not the ability to engage in cognitively complex forms of reasoning."

Several other studies have confirmed this. American psychologist Sylvia Scribner analyzed workers in a dairy who had to fill orders for various products, of varying amounts, from various boxes on shelves. Scribner observed that maximum economy of effort required sophisticated reasoning, again involving nonlinear and interactive effects. But performance was unrelated to IQ score. German psychologist W. Putz-Osterloh and colleagues had subjects take over the management of a small tailor's shop. They had to manipulate a large number of variables involved in the purchasing of raw materials, modification of production capacity, number of shirts produced, amount of profit, and so on. The investigators found no relationship between performance on this system and measured IQ, and concluded that "real" situations involve a complexity of problem structure, and thus of reasoning, not reflected in IQ tests. Other studies, however, have found another factor—self-confidence—to be strongly related to performance in complex problem-solving.

Little wonder that the group from the American Psychological Association could conclude that we "know much less about the forms of intelligence that tests do not easily assess: wisdom, creativity, practical knowledge, social skill, and the like." This would seem to include just about every aspect of human intelligence that really matters.

An Index of Social Class

All this should give us pause to question what IQ tests really measure. On what grounds do they really allocate individuals to this or that rung on the social ladder (often with devastating consequences)? Does IQ really tap an imaginary inner well of intellectual power, or something else? What the IQ movement seems to have been really successful at is the presentation of particular kinds of knowledge or habits of thinking typical of the white middle class, as "general" intelligence.

This bias is obvious with the verbal questions about general knowledge and so on, which make up the bulk of many IQ tests. But it is also easy to show that the reasoning format required by nonverbal tests, like the Raven's matrices, are even more, not less, culture-biased than their verbal counterparts (as we shall see in chapter 7). In other words, what an IQ test measures is assimilation of rather specific, middle-class knowledge and reasoning formats. What is most remarkable is that this success has been sustained even though those forms of knowledge and thinking are unrelated to general cognitive complexity as used and expressed in the practical world.

This is, of course, a suggestion that has been made many times. Perhaps the best source of evidence for it has become known as the "Flynn effect." This is named after American psychologist James Flynn, whose extensive surveys have shown that mean IQ scores of groups all around the world, far from remaining static as the biological theory of g would demand, have risen markedly year after year, decade after decade. In many countries, the increase has been as high as fifteen points over two or three decades. In truth, it has been noted as a problem among IQ test constructors themselves since the 1930s.

Flynn and others suggest (not very convincingly) that the improvement is due to increases in test-taking skills rather than underlying intelligence. But I think the mystery is solved as soon as we drop the illusion of IQ as a measure of general ability and view it, instead, as samples of culturally specific knowledge and reasoning formats. Thus, the

IQs of children and other people rise as soon as they begin to merge culturally with the middle class. The dramatic rises in IQ reported by Flynn have accompanied significant demographic swelling in the numbers of the middle class over that period. The same effect has also been shown in other ways. When members of different social classes in one country migrate to another (as with Japanese to America after World War II) and take up new economic roles, previous IQ "gaps" between them disappear. And, as we shall see further in chapter 3, it has long been known that the IQs of adopted children become much closer to that of their (largely middle-class) adoptive parents than to that of their biological parents, with increases in IQ as high as fifteen points reported.

In view of the transparency of the pretensions of IQ tests, it seems remarkable how many psychologists use them, or cite IQ scores, as if they were getting a purchase on a tangible quantitative variable. Sometimes they will use IQ scores to assure others, in a research design, that intelligence has been controlled for, or matched across groups, as if they really knew what they were controlling or matching for. Similarly, in assessing children's failure at school, an IQ test is often thrown in as an index of general potential. Such acts of faith might explain why psychology remains a backward discipline. Despite the very strong, and often dangerous, claims that emanate from the IQ field, it is not a coherent science.

It is even worse when IQ is used to bolster the kind of claims made by J. Philippe Rushton, and by Herrnstein and Murray, the authors of *The Bell Curve*. Much of this kind of story-making involves rather extreme interpretations of data, and many nonscientific presuppositions. As the group from the American Psychological Association set up to evaluate *The Bell Curve* pointed out, the authors "have gone well beyond the scientific findings, making explicit recommendations on various aspects of public policy." But the depressing aspect about the whole thesis is how little it has changed. As Thomas Bouchard of the University of Minnesota has noted (seemingly with some pride), reviews of Galton's books published more than a century ago could, with a few minor changes, pass for reviews of books published today! This all suggests something (consciously or unconsciously) other than objective science at work.

We will look again at what IQ measures after further discussion of intelligence systems in chapter 5. In the meantime, we should note the rather poor conceptual foundations for carrying out searches for the biological bases of IQ, or even the "genetics" of IQ, which we will discuss in the next chapter.

BIBLIOGRAPHY

Ceci, S. J. 1991. "How Much Does Schooling Influence General Intelligence and Its Cognitive Components? A Reassessment of the Evidence." *Developmental Psychology* 27: 703–22. Critical analysis of the relationship between school attainment and IQ.

Ceci, S. J. and J. K. Liker. 1986. "A Day at the Races: A Study of IQ, Expertise, and Cognitive Complexity." *Journal of Experimental Psychology* 115: 255–66. A classic study of complex cognition in context.

Ericsson, K. A. and N. Charness. 1994. "Expert Performance: Its Structure and Acquisition." *American Psychologist* 49: 725–47. A comprehensive review of what can and cannot be said about the causes of expert performance and "giftedness."

Flynn, J. R. 1987. "Massive IQ Gains in 14 Nations: What IQ Tests Really Measure." *Psychological Bulletin* 101: 171–91. A concise summary of these startling findings.

Galton, F. 1990 (1869). *Hereditary Genius: An Inquiry into Its Laws and Consequences.* [No city]: Peter Smith. Pretty much where the twentieth-century IQ debate has its origins.

Gould, S. J. 1996 (rev. ed.). *The Mismeasure of Man.* New York: Norton. Still a classic for its readability regarding the "reification" and mystification of intelligence in the work of Spearman and the "factor analysis" movement.

Herrnstein, R. J. and C. Murray. 1994. *The Bell Curve: Intelligence and Class Structure in American Life.* New York: Free Press. This is the book that caused the most recent controversy about IQ, and social class and "race" differences.

Lowe, R. 1980. "Eugenics and Education: A Note on the Origins of the Intelligence Testing Movement in England." *Educational Studies* 6: 1–8. Detailed, well-documented, yet highly readable history of the connection between IQ testing and the eugenics movement in England.

Neisser, U. et al. 1996. "Intelligence: Knowns and Unknowns." *American Psychologist* 51: 77–101. This is the report of the task force (the American Psychological Association group) set up to assess "knowns and unknowns" in the intelligence debate in the wake of the publication of *The Bell Curve.* I comment on it in various places (sometimes critically) in the present work.

Sternberg, R. J. and C. A. Berg. 1986. "Quantitative Integration—Definitions of Intelligence: A Comparison of the 1921 and 1986 symposia. In D. K. Detterman and R. J. Sternberg, eds., *What Is Intelligence? Contemporary Viewpoints on Its Nature and Definition.* Norwood, N.J.: Norwood. Psychologists' conflicting definitions of intelligence.

Sternberg, R. J. and P. A. Frensch, eds. 1991. *Complex Problem Solving: Principles and Mechanisms.* Mahwah, N.J.: Laurence Erlbaum. Includes articles analyzing relationships between IQ and performance in situations requiring complex thinking.

Wolf, T. H. 1973. *Alfred Binet.* Chicago: University of Chicago Press. A sympathetic biography of the founder of the modern IQ test, including his views on its subsequent uses.

Zendeland, L. 1998. *Measuring Minds: Henry H. Goddard and the Origins of the American Intelligence Testing Movement.* New York: Cambridge University Press. A complete, richly documented history of the first applications of the Binet test, and its long-term consequences.

Does Biology Hold the Key?

Searching Biology for Intelligence

Despite the obvious flaws in the IQ view of intelligence, the supporters of the test have continued to insist that it taps some fundamental
biological entity, whatever that means. Cyril Burt, following Charles
Spearman, arrived at his famous definition of IQ as a measure of
innate, general, cognitive ability. Philip Vernon followed suit in the
1960s with a distinction between the psychological expression of intelligence (intelligence B) and its underlying form and variation (intelligence A), which are biologically determined. In the 1980s, H. J.
Eysenck defined intelligence A as "the biological fundament of cognitive processing, genetically based (perhaps entirely so), and responsible for individual differences in intellectual competence." These are
the definitions of intelligence that have been fed to legions of psychology students and trainee teachers.

Following from these convictions have been a series of attempts to
demonstrate that more or less good intelligence (or rather, IQ) is due
to variation in quite simple aspects of the hardware of the brain; that
this, in turn, is due to more or less "good" genes; or that it manifests
itself as a genetically determined structure (or series of structures) fashioned by evolution and coded in DNA. Accordingly, many contemporary psychologists now claim to be founders of a more biological,
and thus scientific, "evolutionary psychology." In this chapter we shall
examine these efforts to reduce human intelligence—both its form
and its variation—to a "biological" character, over which its psycho-

logical expression is but a thin veneer. It is worth starting with a strand already mentioned in chapter 2 and initiated by Galton more than a century ago.

Smart Means Fast

As described in chapter 2, Galton's tests of sensorimotor powers as a test of intelligence failed, and for several decades the idea that "smart means fast" was dropped from theoretical consideration in intelligence. Reaction time, however, has recently made a major comeback, some writers describing it as a reflection of "processing speed" or "neural efficiency" on which intelligence itself depends. The modern technique is to have people hold a bar or button down on a console until a light comes on in front of another button, which they then have to press as quickly as possible. Most of the recent excitement stems from studies of W. Hick, the British psychologist, in the 1950s, who found not only that responses were delayed as the number of buttons increased, but the degree of delay differed for different individuals. The excitement was created when it was suggested that the delay was less for those with higher IQs; in other words, a moderate correlation between IQ and this change in "choice reaction time" over increasing choices was reported.

However, initial claims that a biological measure of intelligence had been found were dashed when it was shown that reaction time is far from being a simple, pure response. It involves a number of complex cognitive and noncognitive factors, including understanding instructions, familiarity with the equipment, motivation in the task, sensory acuity, different strategies in various aspects of response selection and construction, and so on. As the task force set up by the American Psychological Association explained, "it is difficult to make theoretical sense of the overall pattern of correlations, and the results are still hard to interpret."

Other searches for simple measures of speed that would explain variance in IQ have produced similar confusion. For example, there have been attempts to demonstrate that IQ is based on "neural efficiency" by measuring peripheral nerve conduction velocity in nerves in the arm (with the obvious, if crude, assumption that this will be the same as the speed of processing in the head, which, in turn, might dictate the level of intelligence). One study reported a correlation of 0.45 between nerve conduction velocity and IQ. However, the simple idea that nerve conduction velocity reflects faster processing which causes the variation in IQ was complicated when it turned out that there was little, if any, association between nerve conduction velocity and reaction time.

As a result of these disappointments with reaction time, IQ theorists have turned to "inspection time." This is the length of time required to discriminate accurately between, say, two lines of different length presented for brief, but slowly increasing, periods on a screen facing the subject. The time required for a subject to judge whether the lines are the same length is said to reflect his/her processing speed. A number of studies have reported moderate correlations between inspection time and IQ in adults and children. Although these results have also been questioned on a variety of methodological grounds, correlations in the range 0.3 to 0.5 have generally been accepted.

The question is, once again, what are we to make of this kind of correlation? We don't know what differences in either of these measures are really differences in, and we have no idea what mediates the correlation, if it exists. Advocates of the approach, such as Ian Deary and Peter Caryl of Edinburgh University, suggest some "basic processing mechanism"—perhaps "quickness of intake of information"—as the basis of individual differences in both inspection time and IQ. But the association could just as easily be due to confidence, familiarity, motivation, or attentional factors, for example. More generally, I hope you will appreciate the irony in the attempted reduction of a complex regulatory system, which cognitive psychologists are still striving to understand, to a simple speed-of-input variable. An analogy may be the explanation of individual differences in overall physiological metabolism by the rate at which we intake food.

As other advocates, such as Australian psychologists C. Stough and T. Nettlebeck, have had to admit, such correlations do not provide us with any explanation for the underlying factors mediating this association. The task force from the American Psychological Association seems to share this pessimism about the whole approach. In any case,

psychologist Michael Howe has argued that many of the reported correlations are inflated by dubious adjustments, especially the widespread (but obviously agreeable) habit of "correcting for unreliability,"
which invariably boosts the correlations reported. He argues that there
are no scientific foundations for the practice, except for making otherwise indifferent results look good.

Behind all these methodological problems, of course, is that of the
validity of the underlying assumption: how credible is the idea that
individual differences in all our intelligent learning, problem-solving,
decision-making, and so on, are determined by elementary speed variables? Any such credibility would seem to entail a remarkable paradox:
if basic physiological speed parameters are at the root of individual differences in intelligence, then we would expect them to be far superior
in humans compared with all other species. On the contrary, it seems
highly unlikely that there is a huge differential in favor of humans in
such simple physiological variables.

Electroencephalograms and Evoked Potentials

Patterns of electrical activity of the brain obtained from electrodes
placed at various places on the scalp, or electroencephalograms (EEGs)
as they are known, were studied in the first half of the twentieth century. Most recent work, however, has been based on so-called evoked
potentials. The presentation of a light flash or an audible click evokes
a wave of electrical activity that is superimposed over the background
EEG activity. In an article in the journal *Nature* in 1969, which largely
inspired the more recent studies, German psychologists J. P. Ertl and
E. Schaefer claimed that "evoked potentials . . . could be the key to
understanding the biological substrate of individual differences in
behavioural intelligence."

Early results suggestive of an association between event-related
EEG measures and IQ produced much hyperbole. For example,
Eysenck argued that "we have come quite close to the physiological
measurement of the [genes] underlying the IQ test results on which we
have had to rely so far." And it was soon being suggested that conventional IQ testing could now be replaced by this "more objective"

approach. Again, however, this initial excitement was soon tempered when a host of unreliabilities came to light (summarized by Cambridge psychologist Neil Mackintosh in 1986), as well as uncertainties about the precise interpretation of an EEG wave, and the role of contextual, attentional, and personality factors. In addition, some studies have produced results opposite to those expected. In a review paper in 1997, Deary and Caryl concluded that "as the task chosen can evidently determine the precise relationship between IQ and brain activity, the vision that conventional IQ tests would in future be replaced by electronics seems a pipe-dream."

Brain Size and Metabolism

For a century or more there have been dreams of "proving" the superior intelligence of this or that "race," class, sex, or other group by demonstrating a relation between brain size and IQ. Earlier studies used measures of head size and have been strongly discredited on several grounds. For example, it turns out that there is little relationship between head size and brain size. More recent studies, using magnetic resonance imaging of the brain, which is assumed to be more accurate, have suggested some association between brain size and IQ. One recent study by American psychologist Laura Flashman and colleagues suggested a small correlation of around 0.25. Again, though, we have to remember that there are many possible ways in which such a correlation could be brought about, other than a direct causal relationship. It could be, for example, that those individuals who have the culturally specific knowledge and self-confidence required for IQ tests are also those who are better nourished, or are taller and heavier, in turn because they come from better socioeconomic backgrounds. After all, a correlation between body size and brain size of 0.5 to 0.6 has long been known, and there have also been reports of an association between self-confidence and test performance. In addition, we have to remember that individuals with very small brains (less than half the average size) have been identified, and are otherwise perfectly normal. And female brains are about 15 percent smaller than those of males, with no apparent difference in intelligence.

Other attempts to demonstrate that IQ is a reflection of neural efficiency have included monitoring people's brains for glucose metabolism using another scanning technique, positron emission tomography (PET scans), during IQ test performance. In one study, individuals who scored higher on Raven's matrices tended to burn glucose at a lower rate, suggesting that they were using less effort. Again, though, neither a cause nor its direction are identified by such correlations: using less glucose could be a consequence of a developed skill in answering familiar questions in a familiar format.

In short, although there have been numerous attempts to anchor IQ variation in differences in biological efficiency, they have offered nothing substantive in terms of causal relations between biological and cerebral variables and IQ. Again, one of the main problems is that the data consist merely of correlations which are open to myriad interpretations. Even reviewers such as Deary and Caryl, who support the search, have concluded in a recent review in the journal *Trends in Neuroscience* that, even where associations are relatively well established, we are far from clear about what's causing them.

Genetic Variation and Intelligence

From Galton to contemporary studies, it has been difficult to find a general biological intelligence in physiological or information-processing parameters. Since the 1930s, however, investigators have attempted to look even deeper—at the genes themselves—for a biological substrate for IQ variation. Indeed, they have called themselves "behavior geneticists," with the implication that they are genuinely describing the genes underlying the development of, and variation in, intelligence, much as their counterparts are doing (with much greater scientific precision) with the behavior of other animals. Again, this search is not concerned with theory about the nature of intelligence as such, but about finding evidence for variation in genes associated with variation in intelligence (almost always conceived and measured as IQ). There have been many motivations for such a search: from proving that cognitively important genetic differences between "races" really do exist (the eugenicists' dream) to the suggestion of the American Psychological Association group that identifying genetic differences may help with intervention strategies, as it has with certain clinical conditions.

Perhaps it should be mentioned at the outset, though, that prominent figures in this debate see this research largely as a matter of confirmation of what is already a "given" of biological laws. As with members of the general public, they see any other possibility as inconceivable. This could hardly have been put more strongly than by the famous British psychiatrist Sir Michael Rutter: "It is manifestly absurd to claim that there is no evidence of genetic influences on variations in cognitive skills. Indeed, it is biologically ridiculous to suppose that there would be no such genetic influence."

Such views might suggest, then, that this section ought to be a brief one (in addition to which, the controversies surrounding these claims have now been discussed many times). However, we shall see that these issues are not as simple as they may seem, and that their implications are deeply controversial. So it does seem important to do at least three things. First, we should briefly review the empirical claims and indicate the rather remarkable assumptions underlying the methods from which data have been derived. Second, we should examine the special, crude kind of genetic theory from which those assumptions are derived. And third, we should highlight the striking self-contradictions that arise from those who use these methods. Finally, and this is something we will take up much more thoroughly in chapter 5, we shall take a look at how the biological laws that are deemed to present such foregone conclusions are rather more complicated than those implied by people such as Michael Rutter, and so much greater circumspection is required. We shall see that it is absurd to suggest that we could, even theoretically, map from arrays of genes to differences in intelligence distinguishing individuals, "races," or other groups.

Empirical Claims

Because it is generally accepted that the development of IQ will be affected by both genes and environments, the problem facing researchers has been how to tell the size of effects of one from the size of effects of the other. Note that we are asking about the size of effects on variation—that is, differences in intelligence—not on its degree in a specific individual. The latter would make no more sense than asking which has the biggest effect on the area of a rectangle, the length or the width? Even so, we can answer the question only if we can control for the effects of one factor while allowing variation due to the other to "show."

This is the reason that twins have been so popular. Try to imagine a population of people as containing a "gene pool," some genes of which (relevant to a chosen trait) may vary from individual to individual, while other genes may not. We don't know what the case is for intelligence (or IQ), but it is known that some traits show little or no genetic variation, while others show a lot. The point is that, whatever the case, identical (monozygotic) twins are derived from the splitting of a single egg in the uterus, and so share all their genes (they have the same genotype for every trait). If they are then reared apart, in different environments, as has sometimes happened in real life, then a comparison of the pairs in a trait, such as IQ, will distinguish the effects of shared genes from the effects of an unshared environment (or so the argument goes).

Several studies of separated identical twins have reported intra-pair correlations of about 0.7 in IQ (remember that a positive correlation can extend from zero to one, which would be the case if a perfect match between pairs was found). Because the unshared environment has depressed the theoretical maximum by only about 30 percent, the complement of 70 percent is thought to be a direct estimate of the heritability of IQ, or the proportion of the total variation in IQ which is accounted for by genetic variation.

However, separated identical twins are rather rare, so numbers are small, and they are not usually representative of the general population in crucial aspects such as socioeconomic group or ethnic group. Another approach has been to compare these monozygotic twins that were reared together with dizygotic (nonidentical) twins reared together. Because dizygotic twins are derived from the fertilization of two separate eggs, each of which has received genes from two different sources, they will share half of the genes that vary for a particular trait.

This means, in other words, that they are no more alike, genetically, than two ordinary siblings. Larger associations in IQ have been found among monozygotic pairs than among dizygotic pairs. Again, this has been taken to suggest that IQ has a substantial heritability.

Another approach has been to see how closely the IQs of adopted children are associated with those of their adoptive parents. This can be compared with the association shown with the IQs of their natural (biological) parents. Because one comparison (child and natural parent) includes shared genes, whereas the other (child and adoptive parent) includes only shared environment (again, so the argument goes), this seems to be another way of potentially telling apart the effects of genes from the effects of environment. In a typical study, the correlation between adopted child and natural parent is higher than that between the adopted child and the adoptive parent (about 0.3 and 0.1, respectively). Yet again, such figures have been taken to suggest a substantial heritability for IQ.

Even overlooking disagreements about what IQ is measuring, the methodological problems with these studies have been discussed numerous times, so let me indicate just a few of them briefly. There has been no good study of separated identical twins in which the twins have been separated at birth and allocated randomly to rearing environments. The so-called "separated twins" used in investigations turn out to have been separated at ages as late as eleven years, and often only into the homes of relatives, in the same neighborhood, attending the same school, playing and/or communicating with each other over periods of many years, and so on.

The main problem with studies of twins reared together is that they don't take account of "treatment effects," in which parents, peers, teachers, and so on treat identical twins more similarly than nonidentical twins, with obvious effects on intra-pair IQ associations. Moreover, in all studies to date, critical biographical data are vague and/or incomplete, so it is impossible to check on degrees of separation, common treatments, and so on. Studies of adopted children have been confounded by "selective placement," in which adoption agencies have tried to match the adopting home to the child's origins (itself a kind of genetic prescience which assumes what the method is trying to prove). The adoptive-child approach has also been severely criticized because it overlooks the many unique social interactions arising in adoptive families (we will look at this more closely in a moment).

In addition, the whole field is compromised by a "make do" empirical culture, involving a hodge-podge of tests, often consisting of estimates of IQ based on people's social class, school performance, or even performance on three or four test items like memory span. In the face of imperfect studies and highly variable results, an increasingly common device has been to lump all results together to arrive at an average for the lot, as if the combination can somehow overcome common flaws.

Finally, the method of reaching heritability estimates from correlations makes a large number of assumptions about genes and about environments, as well as about IQ, which are almost certainly invalid. Because this criticism is the crux of the matter, it is a good place for us to start further examination.

Assumptions Underlying Twin Studies

The interpretation of substantial genetic variability for IQ, from the kinds of correlations just mentioned, is based on certain assumptions: about the nature of IQ; about the nature of genes involved in its manifestation; and about the environment required for its development. Interpretation of results must be influenced by the validity or plausibility of these assumptions. We will look at them as briefly as possible.

IQ as a Quantitative Trait

It is important to remember that the trait (IQ) about which there has been so much nature-versus-nurture debate attracts little in the way of agreed description. We saw in chapter 2 the great difficulties IQ testers have in describing what they are testing (beyond, that is, simple metaphors). We therefore find ourselves in the strange position of being told about the genetics of a trait which cannot be defined!

Be that as it may, behavior geneticists make the assumption that IQ measures a quantitative trait. All this means is that variation in the trait can be treated as increments on a scale, like height or weight. This may sound reasonable enough, but it is important to point out that many traits are not like that. Some, for example, create discrete categories, such as blood groups in humans, or colors of petals in flowers, rather than a continuous scale. Some traits don't normally vary at all within a species; for example, in humans, numbers of teeth, ribs, vertebrae, and a whole host of common physiological processes are the same in everyone. Other traits develop and vary throughout life, within and between individuals, as a direct result of experience; for example, the thousands of different antibodies that humans produce throughout life. As we shall see in the next chapter, some of these traits reflect a remarkable genetic uniformity in species.

Variation in quantitative traits, however, appears to reflect a large number of both genetic and environmental factors (each relatively good or bad) received or experienced by an individual in different combinations. It is also assumed that each factor makes its contribution more or less independently from all others. So the combination each individual gets, and the IQ he or she will end up with, is a kind of natural lottery. This also explains the prominence of the bell curve in debates: there are many more combinations that will produce moderate (near to average) results in the trait, and fewer that will produce either very large or small values (figure 3.1). So, in deciding that IQ is a quantitative trait, investigators are making big assumptions about its genetic and environmental background.

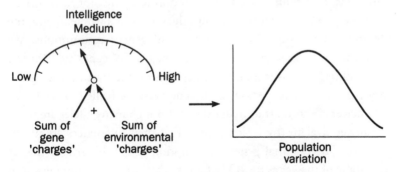

FIGURE 3.1. Behavior-genetic model of the summation of additive gene and environmental unit charges, and thus formation of a normal population distribution of intelligence.

As we saw in the previous chapter, there are many grounds for doubting that human intelligence is just another quantitative trait. It is true that distributions of IQ test scores display many of the properties of such a trait, but these properties—like the bell curve—have been deliberately built in to the tests and are not a neutral consequence of it. Moreover, adaptable traits, such as immune systems, which are active in critical adjustments to environmental changes, can not be treated as degrees on a scale in that way. Ironically, some geneticists have suggested that quantitative traits, as more or less random products of genetic and environmental effects, are "evolutionary dross," so unimportant to survival that the species as a whole is largely unaffected by their variabilities, and so they are allowed to persist. So why do IQ theorists persist with it? The main reason seems to be the statistical and other convenience it affords investigators, and we shall discuss this further here.

The Genes

When investigators use twin studies and adoption studies in connection with IQ, they (consciously or unconsciously) adopt crucial assumptions about the nature of genes and genetic variation, as we have just seen. So let us now look at these more closely.

In fact, we know nothing about the genes involved in intelligence, which is hardly surprising when investigators are not at all clear about what intelligence is. All that we know for sure is that rare changes, or mutations, in certain single genes can drastically disrupt intelligence, by virtue of the fact that they disrupt the system as a whole. Again, the method depends on making a number of genetic assumptions. We know that, in species generally, the vast majority of genes are identical across all individuals (though this depends on the trait). Other genes vary from person to person, and some versions (called alleles) may have better effects on trait than others, although many alleles are functionally equivalent (by which I mean that their differences are unimportant). The essential genetic assumption of IQ theorists is that at least some of the genes for IQ (whatever they are) vary from person to person, and that these can be treated as if they are positive or negative "charges," each affecting IQ independently of all other genes. Like adding the total of several dice, their values just add together, to arrive at an overall "genetic charge" for each individual. The American psychologist Robert Plomin, who frequently uses the model, suggests that "high IQ will develop only if an individual has most of the positive alleles and few of the negative alleles for high IQ."

This, then, is how we arrive at the prediction of a sliding scale of genetic resemblance corresponding with family relatedness. By this simple additive model, pairs of identical twins will share all their genes relevant to IQ development, whether those genes vary in the population or not. Nonidentical twins will share all the invariant genes, of course, but only half of those which vary. The same applies to any pairs of ordinary siblings, and parent-child pairs. On the other hand, pairs of unrelated individuals would be expected to share none of the variable genes. This is the simple calculus that leads to the "expected" correlations in IQ scores: monozygotic (identical) twins, close to 1.0; dizygotic (nonidentical) twins (and other close family relatives), close to 0.5; adopted children more associated with the IQ of their natural parents than their adoptive parents; and so on. It is the fact that the observed correlations accord with this pattern of expectations to a large extent that all the debate has been about. It is on the basis of such a "goodness of fit" that a majority of psychologists have been convinced that part of the variation in IQ test scores is associated with genetic differences between individuals.

What is most remarkable about these conclusions is that the assumptions on which the expected correlations are based are most unlikely, and that those who promulgate them know that. They are unlikely for several reasons. First, in Darwin's terms, natural selection means just that: individuals who possess certain variants of a trait (the well-adapted ones) will be more likely to survive and reproduce. That variant, and any alleles associated with it, will then increase in frequency across generations, while others (and their genes) will be eliminated. The simple implication of this is that traits that have been important to survival do not tend to vary (whereas variation for less important traits will persist).

This has been recognized in breeding experiments in animals since the 1930s, and has been confirmed in organisms under natural conditions. Far from being bags of randomly constituted genes, individual members of a species are already genetic clones in all important respects, or are close to being so. This is also recognized by Plomin, who, in his book *Genetics and Experience*, says that "characteristics that have been subject to strong directional selection will not show genetic variability because strong selection exhausts genetic variability. In other words, when genetic variability is found among individuals in our species, it is likely that the trait was not important evolutionarily." He cites language as an "evolutionary important" character, variation in which may have no basis in genetic variation. But, strangely enough, he fails to consider that the same argument could apply to human intelligence.

The ironic message of those who claim to have identified substantial genetic variability underlying IQ, then, is that intelligence is not an important trait! A more likely possibility, I suggest, is that the simple genetic model being used is not very good. Twin and other correlations may match a model of expected correlations, but the latter are hopelessly idealistic. Indeed, we now know that when genes are involved in the development of important traits, they come to interact with each other as a team, usually under the control of regulatory genes. Because of such interactions, variable genes can be used to produce the same outcomes; different outcomes can result from the same genes; and gene products can even be altered "on line," as it were, by developmental mechanisms. These are all ways in which the most appropriate end point—sometimes a single common target, sometimes individually varying—can be achieved. Genes underlying traits that are important to survival are usually far more controlled and regulated than the lottery of the additive/independent model suggests. Such views have been amply reinforced by molecular genetic studies in the past two decades, and we shall look at these in chapter 5.

The Environment

It is also the case that we know little in detail about the environment relevant to the development of high IQ. Developmental psychologists have disagreed for decades about this, and still do. All we know is that variation in certain broad environmental factors, such as housing, neighborhood, number of children in the home, and so on, is associated with variation in IQ. But this tells us little about causation. In a review of *The Bell Curve*, Robert Sternberg stated that the authors "do not have a very good understanding of the factors that affect IQ. Neither does anyone else; most psychologists are more willing to admit to the fact that factors affecting IQ are still poorly understood." Thomas Bouchard, one of the designers of a famous twin study, who is happy to tell us about the effect of genes on IQ, admits that "in spite of years of concerted effort by psychologists, there is very little knowledge about the trait-relevant environments that influence IQ."

This naïveté about what the environment for IQ is, and how it works, has many serious implications for behavior genetic studies. In separated-twin designs, it is a stipulation that individual members of the pairs must be assigned randomly to environments. Attempting to do this deliberately would, of course, be unethical. But in the absence of what, precisely, counts as the environment for IQ, we have no way of knowing whether it has been achieved anyway. In studies comparing pairs of identical with pairs of nonidentical twins, we don't know whether treatment effects are active or not because we don't really know what they might be. In the adopted-child design, although it is assumed that related individuals are being assigned to dissimilar environments, we can not really know whether this is, in fact, the case.

Again, the tendency has been to treat the environment as an anonymous set of independent charges or factors, each ranging from good to bad, with additive effects on what is tacitly viewed as a passive "intelligence." Theodore Wachs of Purdue University has been most critical of this view of the environment and points to ways in which environmental factors themselves interact to produce complex effects. It seems highly likely that such interactions will have confounding effects on adoption and twin designs.

In an article in the journal *Child Development* in 1993, Berkeley psychologist Jacqueline Faye Jackson offers a glimpse of such interactions in adoption studies. For example, it is known that adoptive parents treat their adopted children differently from their normal children, simply because they are adopted. They may even go out of their way to make them different, to allow the child freer development, and have less clear hopes and aspirations for them. Adoptive parents also tend to worry about the personalities, bloodlines, and social histories of the natural parents of their adopted child, and this affects the relationship with the child. In the same journal, developmentalist Diana Baumrind, also of Berkeley, discusses how adolescent adoptees can become highly conscious of their special identity and react negatively to their adoptive parents' standards and values. And in a study reported by Dutch psychologist M. M. Terwogt and colleagues in 1993, it was found that adoptive parents tend to hold stronger beliefs in the influence of heredity than ordinary parents. This, too, may have complex psychological consequences for their adoptive child's development.

In twin studies, it has been shown how dizygotic twins sometimes react against one another, seeking to create a difference or distance. Again, the actions of conscious agents create an environment of development that confounds simplistic designs. There are many grounds for suspecting, then, that the environment of cognitive development is a richly structured milieu, often the creation of conscious agents themselves, rather than positive or negative environmental factors that separately boost or retard a simple quantitative trait. All this suggests that we badly need to get away from the "summation of factors" model of the environment. We will discuss environments at greater length in chapters 6 and 7.

Interactions Between Genes and Environment

As well as gene-gene and environment-environment interactions, we can, of course, also have gene-environment interactions. What this means is that, for example, the effects of one increment of an environmental factor—let's say nutrition—may be disproportionately larger or smaller in the presence of one set of genes than another. Or, conversely, different sets of genes will react in different ways to the same environment. Again, the behavior genetic methods have to assume that no such interactions exist for IQ: only the sum total of gene and environment charges counts in determining the outcome in the trait. In animal and plant breeding studies, however, interactions between genes and the environment are known to be widespread. The developmental biologist Stephen Stearns of the University of Basel has noted that "almost every study that has looked for them has found them."

Note that gene-environment interactions will invalidate the simple additive model as well as expected correlations derived from it. For example, such interactions may make a correlation for dizygotic twins in a character of around 1.0 a not-unreasonable expectation (the same as for monozygotic twins). Or the interactions may work the other way and make the expected correlations more divergent. By choosing different assumptions, we can play this kind of game all day. Because investigators have identified neither genotypes nor environments for IQ (the whole argument is about simple models), it's impossible to check which assumptions are valid and which are not.

But we can be fairly clear that "secondary" gene-environment interactions are common. This happens when differences in a trait that we know are associated with different genes create interactions among people which, in turn, create differences in something else, which may not be associated with different genes. An example may make this clearer. A number of researchers have now demonstrated how facial and other physical appearances, which are likely to be associated with particular genes, significantly affect how children are treated by other people, the intelligence attributed to these children, and so on, and that this affects their self-concept and subsequent cognitive development. A similar interaction has long been known to apply with physical height, as taller people are often attributed with more intelligence.

Keeping Quiet About Interactions

The implications of all these interactions for the behavior genetics of IQ is quite simple. Although twin and adopted-child correlations fit a simple genetic model, the model is of dubious validity. The result is that we do not know what the expected correlations should be. Dizygotic twins may be actually or effectively as genetically alike as monozygotic twins for the character in question. This is most obvious for characters such as numbers of teeth, ribs, vertebrae, and so on. But it is now well known that humans share the vast majority of their genes (itself a logical consequence of natural selection). Where there is allelic variation, it has been shown that the different versions are often just as "good" as each other. For evolved characters, genetic and other interactions can use different genetic material to reach equivalent developmental end points. In addition, the higher correlations reported for monozygotic twins may be entirely due to uncontrolled environmental or secondary interaction effects. There are numerous possible alternative scenarios that are simply never considered. Indeed, we couldn't seriously begin to consider them without much more basic research into the nature of the character (in this case, intelligence) and how it is affected by the environment. To dash into declarations about genetic variation without doing this seems to be scientifically quite illicit.

What is most astonishing is that those who most strongly promulgate the simple genetic model—and the correlations that, they claim, bear it out—have long known that it is unlikely to be valid. In *Genetics and Experience*, for example, Plomin stresses repeatedly the interplay between genes and between genes and environments, but seems to put such thoughts away when discussing twin correlations and heritability estimates. Indeed, there appears to have been a long-standing conspiracy of silence about them. Wachs talks of "decades of attempting to ignore the existence of interactions." The group from the American Psychological Association, assembled to settle this matter, also remind us several times of gene-environment interactions, but then go on to quote heritability estimates computed on the assumption that they don't exist! It is truly worrying that such self-contradiction can exist in any science, and seems to bespeak some deeper imperative, perhaps that of simply using science to "prove" what is already socially assumed or accepted.

The Meaning of Heritability

Most people will have heard or read that this nature-nurture debate is about the heritability of IQ. They will probably interpret this to mean the extent to which IQ is inherited or genetically determined. This is a rather sloppy kind of understanding of heritability that is often, unfortunately, put forward by those who should know better. Heritability, indeed, has a very restricted meaning: it is an estimate of the degree of overall variation in a character in a sample or population that is associated with genetic differences in the same population in the particular environments in which the sample is living at that time.

It is now widely acknowledged that a heritability estimate cannot apply to any other population (in a different set of circumstances) or the same population at a different time, and that it cannot predict the consequences of any kind of intervention. Many who approve of heritability estimates on IQ acknowledge this restriction. Plomin says that a heritability estimate can describe only "what is," rather than predicting "what might be" or "what should be." This is because genes interact with genes, genes interact with environment, and so on, with indeterminate effects, so that experience of environments (and interactions) other than those sampled in the current estimate would be likely to render the latter quite awry. This consensus again indicates how investigators in this area contradict themselves. In attempting to make heritability estimates, they have to assume that there are no interactions. In pointing to the limitations of heritability estimates, they remind us very much of their existence!

I am genuinely puzzled as to why they don't also acknowledge the more serious implications of such recognition: namely, that if heritability cannot predict "what might be"—the effects of altered conditions or possible interventions—it is scientifically useless anyway. This, too, fosters the suspicion that the nature-nurture debate surrounding IQ has more to do with social persuasion than objective science. The American philosopher and psychologist Ned Block has suggested that the only reason psychologists use such estimates is that it adds a veneer of hard science to a discipline much in need of it. But it becomes a more dangerous game when it fosters a social fatalism about people's potential for intelligence, and psychologists suggest that a high heritability for IQ also means that "racial" differences in intelligence (as measured by IQ) might be genetic in origin.

Looking for Genes for IQ

The idea of genes as little charges singularly boosting or depressing IQ has reached an apotheosis in recent years with the inauguration of hugely expensive research programs designed to identify them. The

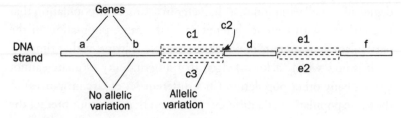

FIGURE 3.2. Genes are arranged as specific sequences on strands of DNA in the chromosomes of body cells. Some genes occur in the population as whole, in various forms, or alleles. For example, gene c has three alleles and e has two (meaning that a given individual will have one or the other).

imagination of the mass media has been fired up with stories of the "first gene for IQ" having already been discovered (described in the journal *Nature* as a possible "turning point in history"), and so on. How could this be so, when IQ theorists agree very little about what IQ measures, and hardly at all about the nature of genetic involvement in it? Here, again, the true picture is not quite the one that is usually presented.

Genes exist as stretches of the chemical DNA which form strands on the chromosomes. The chemical structures of genes can be translated into that of proteins, which are then used as building blocks, regulators, and catalysts in the construction of body parts. In most species, most of these chemical structures (the genes) are identical for all individuals. In others, however, the structure has mutated (either in some current individuals or in a previous generation and transmitted from parents to offspring), so that there are different versions of the gene, called alleles (see figure 3.2). This allelic variation may or may not be reflected in variation in the character.

By using laboratory analytical techniques, it has been possible to identify the loci (or positions on a chromosome) for alleles associated with some traits. This, of course, is relatively easy to achieve when some alleles have obvious pathological consequences: the variants then form a discrete "phenotype" and a one-to-one association can be quickly established. For example, if there are two alleles, e1 and e2, at a locus, and e2 turns out to be much more frequent in the DNA of individuals with a particular undesirable condition, then it might be safe to assume that there was some causal connection, and we would be justified in attempting to describe it further.

Researchers have suggested that a similar method could be used to identify genes underlying normal variation in quantitative traits, such as body size, which involve dozens or even hundreds of genes. Indeed, it has already been used successfully to identify, for example, loci involved in the size of fruit in tomatoes and seeds in beans (both quantitative traits). For other traits, however, the numbers of genes possibly involved, each assumed to have very small effect, make such detective work much more difficult. Those wishing to use the method to identify genes for IQ have suggested a shortcut however. This has been to look only at people at the extremes of the IQ distribution (those with either very high or very low IQ scores), the logic being that alleles positive or negative for IQ are much more likely to show up in such groups. It has therefore been argued that alleles whose frequencies reliably vary across such extremes may be called "genes for IQ."

It is this method of allelic association that Plomin and colleagues, in a heavily funded project involving more than a dozen researchers across five different campuses in the United States and Britain, have been using in the hope of finding the first gene for IQ. They took blood samples from high-IQ and low-IQ subjects (previously tested six- to twelve-year-olds), and submitted them to biochemical laboratories for analysis of the frequencies of alleles at large numbers (up to a hundred) of loci. Where to look was, of course, largely based on guesswork, although their proximity to some genes known to be involved in normal neural functioning was said to be a guide. Likewise, it is not known whether the allelic variants actually have any functional significance, although, as mentioned above, this is usually not the case.

Despite a tremendous amount of scientific and public attention, very little has been found. In the first report of the project in 1995, five alleles showed promising differences in frequencies across the two groups, but these were not replicated in another sample. Other reports have produced promising findings, but, again, these have not been replicated. In a review in 1997, Plomin said: "An analysis of 100 DNA markers has found several suggestive associations with IQ, but none has consistently replicated."

It doesn't surprise me in the slightest that, empirically, this enterprise is turning out to be rather hit-and-miss. What does surprise me is that such a search should be seriously conceived in the first place. One reason for my misgivings, of course, is the idea of trying to identify genes for a character that no one can define. Another is that, even if an allelic association with IQ was found, this would not mean that it was causal. Such a correlation could arise from a number of possible connections other than genes for IQ. How these are to be eliminated is not mentioned.

More important, as mentioned above (and as will be further elaborated in chapter 5), it is rather naive to imagine that genes (at least those related to evolved characters) can be identified as isolated, independent, and additive charges in this way. As just mentioned, the role of a gene in a complex character will vary with the genomic and environmental context. As the Harvard evolutionary geneticist Richard Lewontin has pointed out:

> Genes in populations do not exist in random combinations with other genes. . . . The fitness of a single locus ripped from its interactive context is about as relevant to real problems of evolutionary genetics as the study of the psychology of individuals isolated from their social context is to an understanding of (human) socio-political evolution. In both cases context and interaction are not simply second order effects to be superimposed on a primarily monadic analysis. Context and interaction are of the essence.

Another worry is the extent to which projects of this sort derive much of their credibility from the way that they sensationally reinforce popular notions of biology and psychology derived from social rather than scientific grounds. It is now fairly easy to find accounts in popular magazines and newspapers about scientists having identified genes for IQ, when in fact no such thing has happened. A recent letter in the *New Scientist* (October 30, 1998) carries the idea to its logical and depressing conclusion. The writer looks forward to the day when "a screening test for sperm might one day be used to screen for genetic markers of intelligence, thereby at least increasing our chances of producing brilliant offspring." The sad fatalism that this public dissemination of falsity has cultivated—that the futures of our children will depend on their genes and not on parental or other human action—is seen in the writer's declaration that "I am only sorry that it is not likely to happen soon enough for us."

Note that none of this brings into question the importance of genes in the development of all aspects of the body and mind. We know that variable genes are also associated with variation in many aspects of the development of our bodily parts, such as eye color, hair color, skin pigmentation, and probably height and weight. In addition, we have to remember that many aspects—including, as explained above, those many anatomical features and physiological processes which have been most important for survival in the past—do not vary either genetically or in their functional manifestations.

We also know how some genetic mutations, or changes to the chemical structure that constitutes the genetic "code," have deleterious effects on bodies and minds, and investigation of them is a very worthy enterprise. But this is logically quite different from searching for single alleles involved in normal variation in intelligence. The question at issue here has concerned the extent to which variation in intelligence can be attributed to variation in genes. I hope you are convinced that the most reasonable answer we can offer at the present time is that we have no such knowledge. In chapter 5 we will look at the extent to which it is even a sensible question.

Modules and Constraints

A long-standing approach to human intelligence has consisted of attempts to reduce it to built-in functions by analogy with physical organs. The idea of intellect as a specialized faculty, or a system of several faculties, with innately designed knowledge structures and reasoning processes, has been around since Ancient Greece. The doctrine was reiterated by Descartes and followers in the seventeenth century. In the early nineteenth century, the German physician Franz Joseph

Gall suggested that different individuals developed different mental faculties to different degrees, and that these differences showed in the topography of the skull itself, so that a person's entire psychology could be read by examining bumps on the head.

Although the excessive claims of such phrenologists came to be discredited, the general metaphor of intelligence as a mental organ (or set of such organs), with structure and function encoded in genes, has reemerged strongly in the past two decades. The American linguist Noam Chomsky reinstated the idea in the 1950s when he said of human language, and later of cognition generally, that our usual environmental experience is far too impoverished to account for the development of the rich mental structures we observe in people. Instead, he argued that this complexity is, at least partly, encoded in our genes, just as it seems to be for our bodily organs. In consequence, he suggested that we start to think of a "language faculty," "number faculty," and so on, as "mental organs."

By 1983 the American psychologist Jerry Fodor was calling these organs "modules" to emphasize their self-contained structure. He defined a module as a kind of minicomputer, committed to specific kinds of information. He said they were to be found in specific neuroanatomical structures, in turn encoded by genes. Like many other writers in recent years, cognitive scientist Steven Pinker has suggested: "Our physical organs owe their complex design to the information in the human genome, and so, I believe, do our mental organs." This design information, he argues, has evolved in the genome through natural selection, unconsciously combining, into the production of mental structures, those sets of alleles that enhance survival and reproduction, and eliminating those which diminish it.

These accounts therefore reflect the merging of two more modern metaphors: that of adaptation from evolutionary biology, and that of the computer said to have been shaped by natural selection. This combination of ideas now forms perhaps the most popular theory of human intelligence. We will discuss the computational and psychological aspects of it in greater depth in the next chapter. Here we shall dwell on its evolutionary or biological aspects.

A representative account has been expressed by Harvard educationist Howard Gardner in a series of works from *Frames of Mind: The Theory of Multiple Intelligences* (1984) to *Extraordinary Minds* (1997). Gardner says that there are biologically specified modules that differ in strength or prominence from person to person. They each have their peculiar computational mechanisms, defining the function of the particular "intelligence," and each is based on a distinct neural architecture. He says that the plan for these is in the genome, and that the specified structure will develop even in widely varying circumstances or environmental experiences.

Gardner has appealed to a range of biological sources of evidence to support his thesis. Among these is the fact that brain damage sometimes appears to knock out or disrupt specific intelligences. For example, it has long been known that injury to parts of the left hemisphere of the cerebral cortex can damage language ability, suggesting that cognitive loci are really neurological loci. He claims that neuroscientists have identified neurological units corresponding with distinct cognitive functions. He argues that the fact that all children (and especially many with learning disabilities) excel in one or two domains, yet remain mediocre or downright backward in others, is also evidence of a modular view. He claims that these forms of evidence suggest a biological basis for specialized intelligences, such as linguistic intelligence, logico-mathematical intelligence, spatial intelligence, musical intelligence, and so on. In his more recent work, Gardner has extended his original list of seven intelligences and suggests that the list will grow as further identifications are made.

Many other theorists have come to support this general idea by strong appeals to the logic of evolution and natural selection. For example, American evolutionary psychologists Leda Cosmides and John Tooby argue that the cognitive abilities we now have are actually adaptations that evolved in our hunter-gatherer ancestors more than a million years ago. This has created limitations. Because these abilities were designed to solve specific problem situations prominent then, we can think only in certain limited ways, with more or less limited knowledge structures, today. Pinker has also worried that human intelligence is more adapted to the Bronze Age than the modern age of computers, rapid change, and widespread social networks.

Other recent theorists have argued that, rather than well-structured knowledge and reasoning, or prestructured modules, we are born with certain genetic constraints or predispositions. Modules then develop on the basis of these constraints because of the way the latter direct attention to specific aspects of experience. For example, there may be genetically determined sensitivities in an infant's nervous system which make faces particularly interesting: a "face-processing" module subsequently develops on the basis of this special attention. Modules for reasoning and knowledge still remain genetically determined up to a point. According to cognitive scientist Jeffrey Elman of the University of California and his colleagues: "In this form of nativism . . . the overall structure of the network . . . constrains or determines the kinds of information that can be received, and hence the kinds of problems that can be solved, and the kinds of representations that can subsequently be stored." The whole idea has become very popular in recent years.

Indeed, despite this qualification (and in the absence of clear description of any developmental process), all power still seems to be given to the genes. For example, although a recent book by Elman and his colleagues is called *Rethinking Innateness*, the following few quotes suggest the deterministic role they still accord to the genes: "Genes . . . are, after all, the stuff of which innateness is made"; "genes may sometimes produce morphological and behavioral outcomes with near certainty"; "the gene is like the conductor leading an orchestra"; and, "In our view, the role of the genetic component . . . is to orchestrate these interactions in such a way as to achieve the desired outcome."

Such allocation of creative agency to the genes has led to another major research program in recent years. This has involved the search for modules and constraints in infants soon after their arrival in the world, on the assumption that, if certain knowledge structures and forms of reasoning are present so soon after birth, it proves that they are innate. Accordingly, large numbers of young infants, some only a few minutes old, have been subjected to a wide variety of visual displays, and their cumulative looking times have been taken as indications of their innate cognitive abilities. For example, if they seem to look at cartoon faces longer than at scrambled faces of meaningless patterns, this is taken to indicate that they have an innate face-processing module.

Such results have been taken as firm support for what has been called the "smart infant" thesis: that infants are born with far more intelligence than we have previously credited them with. For example, Elizabeth Spelke of Cornell University has claimed that young infants already appear to have systematic knowledge in physics, psychology, numbers, and geometry. She says this is probably a result of natural selection in the course of biological evolution. In their book *What Infants Know*, published in 1994, French psychologists Jacques Mehler and Pierre Dupoux say that babies seem to come into the world equipped with "a rich model of the world" which is "the cognitive expression of the genetic heritage proper to the human species." In a wide range of other studies, infants have been attributed with innate knowledge or reasoning constraints in just about every domain studied with these methods.

This tendency to attribute knowledge and formative power to the genes has been likened by Susan Oyama, a psychologist at the City University of New York, to the historical tendency in the paternalistic Western world to attribute all creation to a deistic power or god. What is remarkable about the smart-infant evidence is that it is based almost entirely on some form of looking time, whereas the visual acuity of infants under three months of age is actually very poor. In addition, in a typical study, up to 50 percent of results will have to be rejected because of "fussing," a figure that wouldn't surprise most parents. Among results which are marginal rather than categorical (not all infants show the looking preferences), this raises questions about how representative the results really are. Finally, other investigators have suggested that the differences in looking time are actually due to relatively minor sensitivities to perceptual differences in the stimuli, rather than deep cognitive modules in the infants. In 1997, Richard Bogartz, a psychologist at the University of Massachusetts, and his colleagues published a report in the journal *Developmental Psychology* involving a reanalysis of the results of a typical sample of such smart-infant studies. They concluded that all the results can be interpreted in this more basic way. But despite this, theorists like Annette Karmiloff-Smith of the British Medical Research Council readily conclude that, "The neonate and infancy data that are now accumulating serve to suggest that the nativists have won the battle in accounting for the *initial structure* of the human mind" (italics in original).

Problems with Modules and Constraints

We should be clear right away that there is nothing really new about these views. As we saw in chapter 1, thinking as a kind of computation was an idea proposed in the seventeenth century, and the idea of "mental faculties" was popular for a time in the early nineteenth century. What is new is the assumed theoretical underpinning for them provided by evolutionary biology. As in the past, though, there are serious problems with it.

We get a strong whiff of the nature of these problems from the admissions of Gardner himself. In his book *Frames of Mind*, he suggests that "findings from neurobiology" provide us "with a powerful hint about the possible 'natural kinds' of intelligence." At the same time, he warns that we must not think of these intelligences as physically verifiable entities, but only as "potentially useful scientific constructs." We are told of the neurological evidence for modules, but warned that "even the most informed scholars of the nervous system differ about the level of modules that are most useful for various scientific or practical purposes." He is keen to stress how innate knowledge "has been internalized through evolution so that it is now 'pre-wired' in individuals," yet claims that "an intelligence" is a set of problem-solving skills valued by a particular culture.

As with modularists in general, Gardner notes how modules are products of the imagination rather than empirical demonstration. He candidly admits that the selection (or rejection) of a candidate intelligence "is more reminiscent of an artistic judgment than of a scientific assessment," thereby acknowledging the dangers of reification (or attributing real existence to something that doesn't actually exist): "Sympathetic readers, will be likely to think—and fall in the habit of saying—that here we behold the 'linguistic intelligence,' the 'interpersonal intelligence,' or the 'spatial intelligence' at work, and that's that. But it's not. These intelligences are fictions."

This wish to have things both ways by appeals to reified or mystical entities as products of genes is not untypical of evolutionary psychologists. Thus, although the idea of genetic constraints is now very popular, not one such constraint has yet been identified (beyond, that is, hypothetical attentional tendencies or other predispositions). Many will agree with the view of Frank Keil of Cornell University that such constraints act as core principles or structural skeletons of abilities, which still permit considerable developmental flexibility. But this doesn't explain how the historically and contemporary huge variety of forms of intelligence among humans can arise from the same "skeleton." Substantial differences in animal bodies, for example, are not found without underlying differences in the skeleton.

In fact, there is no empirical evidence for such modules or constraints at the level of human intelligence whatsoever. Let me attempt to illustrate this with what must be thought of as the module about which we have the most empirical information (and the one most uniquely human): namely, speech and language. You may remember how Gardner refers to a "linguistic intelligence," which he says corresponds with speech areas in the brain. This idea is based, in turn, on the well-known observation that damage to parts of the left side of the brain often seriously disrupts speech abilities. Pinker also refers to a specific "language bioprogrammer" with a specific "computational architecture." And in his book *How Brains Think*, theoretical neurophysiologist William Calvin of the University of Washington, Seattle, says: "There is, of course, a 'language module' in the brain—located just above the left ear in most of us—and Universal Grammar might be wired into it at birth."

The reason it is worth considering this module in some detail is that it perfectly illustrates the main problem, which is the failure to analyze and sufficiently conceive the "environmental" problem that intelligence has to deal with. It thus also illustrates why this failure necessitates a retreat into the "poverty of the stimulus" argument, and drives us to fall back on to mystical built-in structures.

On the face of it, speech involves the issue of different acoustic signals, which we identify socially and scientifically as phonemes. We can think of these, for present purposes, as the thirty or so vowels and consonants of any human language. These phonemes are combined according to conventional (socially agreed) rules ("morphological" rules) into words, and these, by another set of rules ("syntactic" rules) to form sentences. Together, these form the wonderful expressivity and creativity of speech found in any human language.

Also on the face of it, interpreting speech simply involves the detection of the fixed cues and structural relations in it, by mechanisms which the modularists say are innate. These cues then trigger the retrieval of semantic representations in our memory to recover their meanings. By such means, the acoustic structure of a sound like "boat" becomes clothed in all the meanings associated with it: an image of a boat, the activity of sailing, and so on. When researchers have examined the acoustic stream of speech more closely, however, something other than fixed cues appear. It becomes obvious that the values of a range of variables occur in constantly novel combinations. These include the sound frequencies, their patterns of change over time, the relative timing of onset of sound in the larynx, the places of closure of various parts of the mouth, the identity of the following vowel, the age and sex of the speaker, and so on. There are no simple, tangible cues unique to individual phonemes from which they can be identified. Instead, these identities have to be constructed from the unique pattern of values presented by the many different variables, almost always with reference to their contexts in the wider utterance. What we mentally recognize as a phoneme is not a fixed signal but an abstraction (figure 3.3).

Because the system of speech encoding and decoding cannot depend on discrete cues or fixed structures, there is no compulsion to posit genetic blueprints or preprogrammed modules for them. So how do we end up with "speech areas" in the brain? In their critical survey of the evidence in the *Annual Review of Neuroscience* in 1997, R. H. Fitch of Rutgers University and colleagues say that a speech module is an overattribution of language function to much more general and lower-grade functions. What are claimed to be neural structures evolved for speech are simply structures necessary for making very fine

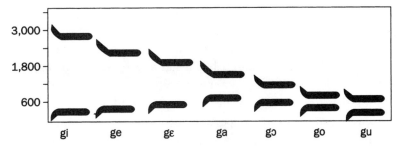

FIGURE 3.3. Although it continues to "sound" like /g/, the actual acoustic energies being transmitted/received vary markedly with context (in this case the following vowel). Each bar shows streams of acoustic energy of different frequencies during the utterance of corresponding consonant-vowel pair. Redrawn from P. Delattre, A. Liberman, and F. S. Cooper, "Acoustic Loci and Transitional Cues for Consonants," *Journal of the Acoustical Society of America* 27 (1955): 769–73.

temporal discriminations. For example, differences of a few milliseconds in a number of variables (together with their contexts) are crucial in distinguishing one phoneme from another, as can be seen in figure 3.3. As crucial as they are for speech, they can be carried out by previously existing sensory and motor capabilities (evolved for dealing with more general, rapid changes in experience) now co-opted for speech processing.

As Fitch and colleagues point out, the traditionally acknowledged speech areas of the brain were identified on the basis of rather gross, imprecise lesions. More recent research, including magnetic resonance imaging, EEG and comparative behavioral analyses, as well as postmortem cerebral examinations, suggests involvement of far more widespread regions in speech processing. All these seem to be regions or architectures already specialized for the processing of rapid acoustic and other temporal information, which have been co-opted and developed for speech processing.

This idea is supported by much other evidence. For example, it has been found that damage to speech areas also creates deficits in processing rapidly changing acoustic stimuli in general, speech or nonspeech. Likewise, the brain areas which, when damaged, appear to have the most serious effects on speech are precisely those that have a general role in the processing of rapidly changing acoustic and other variables. It has been said that the left hemisphere of the brain is specialized for the discrimination of speech sounds, but it has also been shown that this extends to the discrimination of nonspeech sounds when the discrimination depends on rapid acoustic changes. In addition, it has been shown that animals such as apes and dogs can learn to distinguish the phonemes of human speech. And damage to the left side of the brain is known to produce deficits in complex auditory discrimination in nonhuman animals (including their species-specific call signs).

Thus, as Fitch and colleagues conclude:

> Empirical evidence suggests that humans do not, in fact, possess a neural module that is activated only by speech and that makes humans distinct from all other animals. . . . Indeed, the data converge to suggest that speech processing is subserved by neurobiological mechanisms specialised for the representation of temporally complex acoustic signals, regardless of communicative or linguistic relevance in human and non-humans alike.

I suspect it would be fairly easy to extend these, or similar, arguments to all the other candidate modules or innate constraints for human intelligence. As we shall see in chapter 6, there is nothing special about speech having an abstract structure: it is something common to all aspects of human experience (and that our failure to see it is one of the problems). Certainly, scrutiny of the human brain at either a macro or micro level does not indicate the presence of such modules. Indeed, in their review of structures of the brain, Elman and his colleagues concluded that, "There is no evidence that humans have evolved new neuronal types and/or new forms of neural circuitry, new layers, new neurochemicals, or new areas that have no homologue in the primate brain." As we shall see in chapter 8, this conclusion is being increasingly generalized as investigators turn up more and more examples of flexibility, modification, and context sensitivity in what were previously thought to be rigidly specified functions.

In summary, attempts to trawl the crypts of biology for the loci and forms of intelligence, and variation in it, seem simply to have compounded puzzlement. Such is the result of decades of misdirected attention and effort. This is a theme we shall be returning to several times.

BIBLIOGRAPHY

Chomsky, N. 1980. *Rules and Representations*. New York: Columbia University Press. The powerful (and very clearly argued) inspiration for many of the modular theories which followed.

Deary, I. J. and P. G. Caryl. 1997. "Neuroscience and Human Intelligence Differences." *Trends in Neuroscience* 20: 365–71. A broadly supportive review of attempts to reduce intelligence to neurological factors.

Elman, J. et al. 1997. *Rethinking Innateness: A Connectionist Perspective on Development*. Cambridge, Mass.: MIT Press. An attempt to revise the innate modular view by a genetic-constraints-plus-development view, and to illustrate the model with connectionist modeling—although it contains a number of inconsistencies.

Fitch, R. H., S. Miller, and P. Tallal. 1997. "Neurobiology of Speech Perception." *Annual Review of Neuroscience* 20: 331–53. A critical review of data and interpretations surrounding the modular view of speech, and which, in my view, can be theoretically extended to all other hypothetical modules of intelligence.

Gardner, H. 1993 (10th Anniversary Ed.). *Frames of Mind: The Theory of Multiple Intelligences*. New York: Basic Books. One of the first accounts of a modular theory of intelligence.

Hirschfeld, L. A. and S. A. Gelman, eds. 1994. *Mapping the Mind: Domain Specificity in Cognition and Culture*. New York: Cambridge University Press. A collection of papers celebrating the modular view of mind, including one by Leda Cosmides and John Tooby.

Mehler, J. and P. Dupoux. 1994. *What Infants Know: The New Cognitive Science of Early Development*. Malden, Mass.: Blackwell. A concise review of the "smart infant" thesis.

Plomin, R. 1994. *Genetics and Experience: The Interplay Between Nature and Nurture*. Sage Series on Individual Differences and Development, vol. 6. Thousand Oaks, Calif.: Sage. An epitome of the behavior genetic approach in psychology, including mountains of correlations and heritability estimates.

Richardson, K. 1998. *The Origins of Human Potential: Development and Psychology*. New York: Routledge. A wider critique of the assumptions underlying behavior genetic approaches to intelligence.

Rose, S. 1998. *Lifelines: Biology Beyond Determinism*. New York: Oxford University Press. A general critique of reductionism in biology and neuroscience.
Wachs, T. D. 1992. *The Nature of Nurture*. Thousand Oaks, Calif.: Sage. An account of conceptions of the environment in human development, including an appeal for a more interactive view.

4

Computations and Connections

The idea that human intelligence can be described in mechanical terms, as a form of computation, is a very old one. In the seventeenth century, Thomas Hobbes was the first to attempt to materialize psychology when he argued that "Reason . . . is nothing but reckoning." He spoke of "ratiocination" involving "brain tokens," and defined ratiocination as computation. The advent of the modern computer has, since the 1960s, provided a potent metaphor for cognitive science, and renewed thinking about the nature of intelligence. By the 1980s the computational view of the mind had established a more or less standard model of cognition, and now forms perhaps the most popular general framework for the understanding of intelligence. So the first thing I want to do in this chapter is to offer an overview of what is now called the computational model of intelligence.

As we shall see, the view has been tempered recently by much criticism. The idea that the brain is a kind of serial computer, processing parcels of information (called symbols) according to built-in programs, has produced many problems. Rival accounts have attempted to use the brain itself, and its parallel processing among massively interconnected neurons, as the key metaphor. This framework has become known as connectionism, and I will offer an overview of that framework, too. Finally, in several places we will look forward to an analysis of intelligence systems, which will follow in chapters 5 and 6.

The Computational View of Intelligence

What we do with computers is input symbols, such as strings of letters (words) and numbers, and get the machine to process them according to built-in or stored programs (often described as rules) and wait for the output (which may, like some of the programs, be stored in the computer's memory). The processing can be extremely complex, so in that respect at least, what goes on in a computer seems to be a good model of thinking. Likewise, the information and the programs stored seem reasonable models of knowledge and memory.

The general idea also seems to translate to real knowledge and thinking quite readily, as indicated in figure 4.1. Information is input at sensory registers and encoded in some form (usually in symbolic form, such as a letter, word, or image) according to some rules. It is then conveyed to short-term or working memory, where it is analyzed, compared, transformed, or whatever, according to other rules, at least some of which are retrieved from long-term memory. Some of this activity will also involve comparison with, or retrieval of, other information or knowledge representations held in long-term memory. The result of this activity is either some addition to long-term memory or some motor action.

The way that a computational model of intelligence has swept the field is seen in its enshrinement in Steven Pinker's 1997 book *How the Mind Works*. Pinker claims that what goes on in symbol-processing computers offers the best understanding of human intelligence, so I

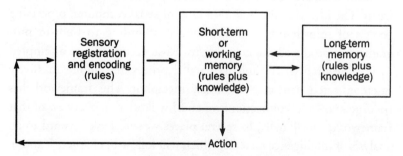

FIGURE 4.1. Overview of processing in a computational model.

will be dipping in to it frequently as a representative of the whole approach. It is an atomistic approach in the sense that it views intelligence as the sum total of a large number of discrete processes into which any complex process in the real world, or thought and action in the mind, can be broken down. Pinker argues that thinking can be broken down into rules, each located in simple units or "demons," each doing a simple task in a reflex-like manner, and handing off output to other demons which behave in a similar reflex manner. The rules are of the simple "If-Then" kind, in the sense that they respond in a fixed and logically defined manner to specific inputs: If a is present, Then b will follow; If the red light is on, Then signal "stop"; If "2 + 2," Then "4," and so on. In most models, some of these If-Then rules will be innate (built in to the module's architecture, as described in chapter 3), and some will be acquired through learning.

Of course, most cognition is much more complex than this. But in every case it can, in theory, be analyzed and described by breaking the task down into progressively simpler components. Eventually, this decomposition will result in tasks that can be handled by the simplest If-Then rules. Each of these corresponds with an equally simple unit, or demon, doing nothing until its preset conditions arise, when it responds in a reflex-like manner, sending a message, which is deposited in memory or is the condition for another such demon to act, and so on. Demons are wired up in series and hierarchies, and the products at one level are always in the form treatable by If-Then rules at the next. By this progressive coordination, demons which separately are very unintelligent can collectively produce all that we recognize as human intelligence. According to Pinker, any problem that can be broken down into a series of logical steps can be solved by a machine that thinks in this way. The computational theory of mind is the hypothesis that intelligence is computation in this sense.

In the illustration Pinker uses, he asks us to imagine knowledge arranged in long-term memory in the form of statements or propositions. These are actually only meaningless tokens, but stand for the state of some aspect of the world and might include propositions of the kind: "A parent of Q" (that is, person A is the parent of person Q); "B sibling of A"; "P is male."

If the system is then asked to confirm or disconfirm that "P uncle of Q," it proceeds as follows. First, we have a demon that searches memory for "P uncle of Q." Because there are no such propositions in the memory at present, another route has to be found. Logically, this means identifying Q's parents' siblings, so the alternative route is broken down into the following steps: "Find Q's parents"; "Find Q's parents' siblings"; "Distinguish as uncles/aunts."

How is this done in the system? Well, we just invent a new demon to perform each new step. Thus the first demon, having been frustrated in its search for the "P uncle of Q" proposition in the memory, now recruits other demons to go through the series of substeps in the new route. These include a demon (triggered by "parent of" relations) that finds Q's parents, and another that finds P's siblings and puts what it finds in a separate list in the short-term memory. Yet another demon sorts them into male and female (aunts and uncles), attaches the relation "uncle of" or "aunt of," and adds the results to long-term memory (because they are new propositions). One of these will, of course, be the crucial proposition "P uncle of Q," which can now be retrieved from long-term memory (by the first demon), and the problem is solved. As Pinker puts it, the machine has "deduced the truth of a statement it has never entertained before."

It sounds simple and highly mechanical, but the apparent capabilities of such symbol systems have been quite impressive. Simulation programs devised in cognitive-science laboratories have achieved processing and outputs that have mimicked, at least to some extent, various aspects of human memory, reasoning, and representation. In the field of machine or artificial intelligence, large numbers of such systems have now been implemented in problem-solving devices, carrying out routine sorting, screening, and diagnostic tasks, or performing in games such as chess. Allan Newell, a computer scientist at Carnegie-Mellon University, has argued that any system of general intelligence can be reduced to such a physical symbol system.

Problems with the Computational View

At one level, it is difficult to dismiss entirely the view that thinking is a kind of computation (even a child will say that solving a problem consists of "working it out"). But the process of logical decomposition of thinking and acting into mini-thoughts and mini-acts on symbols in order to describe how we achieve whole acts and whole thoughts creates many practical and theoretical problems.

We see some of the problems immediately in Pinker's example just described. The illustration is an object lesson in "ad hocery." Demons "just have" all the functions required; whenever a logical function is identified, we just invent a new demon to carry it out. Reasoning crucially starts with the propositions in memory, but we are given no indication of how they got there in the first place. The account is just a plausible story based on logical decomposition of the task. In reality, it has been done by the modeler (Pinker) breaking down the end point ("P uncle of Q") into tacit subpropositions and then inventing If-Then demons to sort, label, and retrieve them. Needless to say, although natural selection or evolution is appealed to as a kind of blanket guarantee of the origins of such demons, of the way they are wired up, in real brains, no causal pathway is ever described. This already implies a number of other crucial problems with the account.

The Symbol-Grounding Problem

When we key things in at the computer keyboard, they are symbols that are already meaningful to somebody (hopefully including ourselves). They already have some connection with the outside world. A major question—known as the symbol-grounding problem—is how this can happen in the kind of computational model described here.

A famous parody of the problem, suggested by Berkeley philosopher John Searle, is that of a Chinese room in which Chinese symbols are processed by a "blind" internal operator. This operator acts only according to a list of If-Then instructions, and knows nothing of the meanings of the propositions received from, or produced for, the outside world: If (receive patterns A), Then (output pattern X), and so on. Searle says that the person (or program) inside is always missing intentionality, so such intentionality must always come from somewhere else; hence the computational model is inadequate. There has still been no satisfactory answer to this problem (although argument continues).

In the same vein, it is certainly of great convenience to have representations of the outside world fashioned and stored in memory as ready-to-use statements or propositions. But this, too, seems to be more an act of faith than serious psychology. As Lawrence Barsalou of the University of Chicago has pointed out, we have no plausible accounts of how propositional representations arise in the cognitive system, either innately or through experience; there is no direct evidence for them; and we have no account of how propositions relate to the outside world. Therefore, we should at least be somewhat skeptical of them and continue looking for alternative models.

Discrete Cues and Fixed Relations

Our (human) relationship with the outside world is not that of a computer with its keyboard, which is based on a reliable, one-to-one association between signal and meaning. Most experience (as we saw with the issue of speech in the previous chapter) is such as to frequently require interpretation of inputs in the context of the whole, rather than as discrete symbolic triggers. Although we can logically decompose problems into simple, repeatable steps, each based on a condition-action (or If-Then) rule, it is easy to show that the real world is not like that.

We don't experience even the world of objects as a series of static images. For example, a chair in front of you might create a stable image in your mind, but this is a result of your cognition, and not a direct copy of what's out there. It is more than likely that the real chair is at a novel combination of distance, orientation, and angle of vision. The problem may be confounded by the fact that the chair may sit in among numerous other objects, all showing similar cues or features (edges, corners, and so on), and you have to bind the collection that belongs to one away from the collections that belong to others. You may also be moving around it, so the input is constantly changing, with some parts now being occluded, now coming into vision. Even standing still and staring at it doesn't produce a stable input: what you will be unaware of are tiny oscillations of the eyeballs in their sockets, themselves creating ever variable images on the retina at the back of the eye.

It is one thing to invent a computational system to deal with such a world when we key it in at a keyboard as "chair back," "table leg," and so on. But how do our brains know that's what they are, from novel perceptual images, in the first place? In fact, no one really knows how the brain maintains such object constancy in the face of changing sensory information (and it remains a fundamental and pervasive problem in psychology). Because each presentation of an object, and of each of its featural cues, is virtually unique, we would have to envision a computational system for dealing with it made up of millions and millions of If-Then demons, each of which may be used only a few times throughout life.

Vision is not unique in this. The worlds of touch and kinesthesia (in our muscles, tendons, and joints), of sound (including the acoustic stream of speech), and of all the other senses are dynamically changing. Those worlds, and their features, are virtually never experienced in the static canonical views required by symbol-crunching machines. There are few features in the world that are stable and repetitive in the way that reflex demons would need them to be. Each input to a real, organic (as opposed to mechanical) system is unique.

Yet our intelligence has an astonishing ability to recover meaningful information even when such inputs are highly skimpy, fragmented, or degenerate. Consider, for example, so-called point-light stimuli. A classic example, the point-light walker, is shown in figure 4.2. It has been known for decades that just a few points of light can produce very rapid recognition of an object if the points are moving in a coordinated fashion (but not if they are static). Yet they contain no features or symbols of the sort required by If-Then demons. Instead, recognition seems to depend closely on the structural complexity of information between the points as a whole.

Just as sensory information doesn't easily break down into the discrete forms required by the computational model, typical everyday human problems do not break down into repeatable, discrete steps. Consider a piece of meat (or some other food) you've had for four days, and you're wondering whether it is edible. You might decide that four days is too long, so it must be thrown out. But then you think, "Well, it's been in the fridge most of the time." And then you consider that it's for you (a fit and healthy, toxin-resistant adult), and not for a

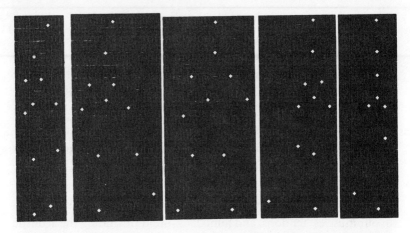

FIGURE 4.2. A sequence of images from a point-light walker. A person walking is quickly recognized from such a display (the images have been spaced out for this figure). *Source:* D. S. Webster and K. Richardson.

young child, or for an aged or infirm relative. So you decide it's safe to eat. Consider another typical problem, such as deciding whether to travel to Boston by train. Train travel is relaxing and relatively fast. But then you consider punctuality, reliability, and expense, and possible alternatives, like traveling by car or airplane, and you think of their relative advantages and disadvantages.

In each of these situations it would be bizarre to suggest that your deliberations consist of the grinding of fixed condition-action demons in your head: If (older than three days), Then (discard food); If (destination Boston), Then (go by train). Instead you decide how one variable (such as the age of the food) predicts another (whether it's edible) by bringing in other variables from your background knowledge and judging how they modify the prediction. A number of psychologists have pointed out how everyday knowledge and cognition are highly conditional in this sense, and how such conditioning can run very deep (in the sense of the number of times you can refer to another factor to condition the effects of the one before it, and thus the final prediction). Real-life cognition is like this because that's the way experience is in complex, ever-changing environments. The assumption that the environments or experience of people can be broken down into simple If-Then steps seems far too idealistic. We will look again at the structure of complex environments in chapters 5 to 7.

Limitations of Machine Intelligence

In these and many other analyses of the real structure of experience, it becomes clear that the computational crunching of ready-made symbols is applicable only to an ideal, ready-made world, one which a machine of intelligence far more modest than that of humans can cope with. Perhaps it should not be surprising that, from both practical and theoretical points of view, the results of applications of the computational view—that is, artificial intelligence—have been distinctly mixed. As problem-solving aids, some have worked and some haven't. The ways in which they haven't are interesting because they have further highlighted aspects of human intelligence that are at odds with the computational view.

One indication of this is the way that the search for a computational general problem-solver has gradually given way to the design of more specifically targeted "expert systems." Artificial intelligence programs work best with simple, discrete signs, such as medical symptoms or numbers, which can be (socially) demarcated before being input at a keyboard and operated on. It has been much more difficult to devise programs that work well with fuzzy or widely variable inputs, such as a moving image or human speech. Input therefore has to be put into humanly intelligible form first, before the computational intelligence can get to work; no regard has been given to how this happens in the real cognitive system. In view of this remoteness from the demands facing a real intelligent system, it is, perhaps, not surprising that Pinker is puzzled about why "a computer finds it more difficult to remember the gist of 'Little Red Riding Hood' than to remember twenty-digit numbers; you find it more difficult to remember the number than the gist."

In addition, because the computationalist view, with its reflex demons, is best suited to a stable world of fixed chunks of information, it has always had difficulty in explaining how new rules may be acquired, except in an ad hoc, unprincipled way. Yet in a dynamic, changeable world, especially one that changes by its very actions upon it, an intelligent system will have to learn new rules all the time. A characteristic problem in artificial intelligence research is that the sets of existing rules are often unable to accommodate new rules, making the system very brittle and creating bottlenecks. This, in turn, requires continuing intervention on the part of the programmer, which is hardly characteristic of a system that can learn automatically.

Finally, it is worth emphasizing how this mind machine of the computationalist is a slavishly obedient system, adapted to fixed environmental requirements, and so is not an active intelligence at all. It is a "decognitivized" machine, an idealistic parallel of the cognitively disenfranchised production worker, or "hierarchies of domination along mechanical lines," as British psychologist John Shotter put it. It is striking, indeed, how Pinker's descriptions of hierarchies of dumb demons parallel ideal, rigid production relations in the factory, mill, or office. A quite different system therefore has to be proposed, and we shall turn to that problem in chapters 5 and 6. In the meantime, let us return to the main alternative to symbolic computationalism as an account of intelligence that has become popular in recent years.

Connectionism

The metaphor of ranks of dumb demons acting on discrete symbols and propositions has been unhelpful with regard to real (as opposed to artificial) intelligence. Investigators have therefore sought new metaphors to help overcome such problems. Increasingly clear and impressive accounts of the composition of the brain itself, with its richly interconnected network of neurons, seemed by the 1980s to offer just such a framework. Since about the 1980s, the computer metaphor of intelligence has been rivaled by a brain metaphor, in which learning and cognition take place through "neural networks" put together on a computer, in an approach generally known as connectionism.

Modern connectionist models consist of processing units that are dumb, like Pinker's demons, but are even less semantically wise. The inputs to which they respond are much less coherent than whole symbols, and, individually, these units do something much less structured and useful than even a simple If-Then rule dictates. Instead, they are responsive to combinations of impulses from large numbers of other units (some of which may be sensory receptors). If the aggregate of impulses from those inputs exceeds a certain threshold in the receiving unit, the latter will fire, or transmit an impulse, in turn. Now this impulse goes to another unit receiving large numbers of inputs . . . and so on. The units can be seen as junctions-boxes, in a widespread network of interconnections between units, some of them connected to receptors responsive to "meaningless" features, such as lines, colors, or locations, others producing outputs of various kinds.

But this is not entirely a recipe for myriad impulses tearing chaotically around a network simultaneously and in random directions. Instead, the overall pattern of firing is constrained by past experiences of impulses. In actuality, each connection on a unit is modified according to the number of times it has been used before: the more it is used, the easier it is for the next impulse to get through and help fire the receiving unit. In this way, whole patterns of inputs can be represented in the pattern of modified connections across numerous units. A visual pattern appears in the outside world. Different input units are sensitive to different aspects or features of the pattern, and fire accordingly. Each impulse is sent to numerous other units to which the input units are connected, which fire in turn if the aggregate of inputs is strong enough. After several repeated presentations, this becomes increasingly likely because the connections are modified and the different features thus associated. The original pattern is then said to be represented in the system, not as a direct copy or imprint but as a pattern of weights on the connections between the units.

Two properties of connectionist networks stand out. First, meaning is not found in the processing of discrete tokens or symbols, each corresponding with some external state (for example, features, whole object images, or language-like propositions). Rather, meaning is found in a pattern of activity across the network as a whole. Intelligent activity is said to be distributed rather than localized. The second outstanding property is that the system "learns" in a way rather more obvious than in symbolic computational models: by modification of connection weights.

This arrangement appears, at first glance, to get us off a number of theoretical hooks. And it appears to furnish several other advantages as well. One is that an input image, bearing only a few of the features of a previously experienced object (perhaps a partly occluded head of a dog sticking out from behind a wall), will still result in an activation of the entire representation, with legs, tail, fur, and all (and, indeed, all the many other things that have been associated in your past experience of a dog). This means obtaining predictability from partial, changeable, or otherwise fuzzy input (which is certainly a reflection of everyday experience, as we saw in chapter 3).

Another impressive property is that the same set of units and connections can, in effect, represent a number of different concepts. Because these reside in a specific pattern of connection weights, several of them can be overlain in the same network without mutual interference. There is therefore no need for the proliferation of processing units or vast expansion of storage space as our experience increases, and as might be required in symbolic models. Moreover, the arrangement implies that damage to some of the output units will result in incremental, or "graceful," degeneration of representation and activation, rather than a cataclysmic breakdown as in a traditional symbolic processor. This property, too, is closer to what happens with real injury to real brains.

The successful simulations of the results of simple human problem-solving are what have attracted most attention and credence. A simple network can simulate the results of some experiments in concept formation, for example. If the sets of feature values (small eyes, large nose, medium mouth, and so on) of the schematic faces shown in figure 4.3 are fed in, in succession, at the input units, they may associate in such a way as to exhibit a "prototypic" representation at the output units. That is, the prototype will come to evoke the highest level of activity in output units, even though it has never actually been experienced. This mimics what is thought (rightly or wrongly) to happen with humans given such experiences.

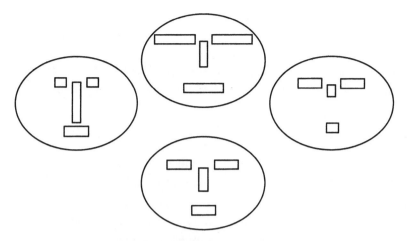

FIGURE 4.3. After experience with a range of faces of different combinations of feature values, a prototype (bottom center) is said to form in representation. If the values are keyed into a connectionist network, the prototype comes to create the biggest activity in the output unit.

Such association may happen automatically with a simple network, but this is rare. In reality, so-called hidden units are needed between inputs and outputs in order to partially associate inputs in some way, before they are sent on to output units. In addition, the network usually needs to be "trained"; for example, the deviation between the network's strength of response to a specific pattern and that which the investigator wants or expects is fed back into the hidden units. The latters' connection weights are then adjusted by the (very human) modeler to bring the output more into line the next time that pattern is presented. Such adjustments progressively bring the network to work in the way required, even for novel inputs.

By such means, complex patterns of weights can form, which, it is claimed, approximate complex, abstract rules, such as those involved in categorization, object recognition, and speech. Networks can learn "words," so as to recognize distorted or smudged versions, and the difficult (irregular) past tense forms of English verbs (recognizing "dug," but not "digged"). That is, they may exhibit "emergent properties" at least superficially resembling those actually implemented in real brains or minds, purely on the basis of strengths of association. Domains (or domain-specific knowledge structures) can be constructed by such emergent activities, offering an alternative account of the origin of specialized cognitive modules.

This is a very skimpy look at the principles of connectionism, and those who want to do it more justice are urged to consult some of the standard texts mentioned in the bibliography at the end of this chapter. But is it the breakthrough in the understanding of human intelligence we have long sought? Its supporters claim that connectionist models are more than electronic toys for doing specific tasks—cartoon brains, consisting of cartoon neurons, with cartoon synapses, as American cognitive scientists Stephen Hanson and David Burr called them—but accurate models of human intelligence. But we should be aware of some of the many problems about the enterprise which throw doubts onto such claims.

Criticisms of Connectionism

The first problem is that, although networks can be shown to mimic some aspects of human cognition, this doesn't just happen spontaneously, within the network itself, on the basis of the naturalistic experiences typical of real brains or minds. The typical connectionist network has to be set up by the modeler on the basis of numerous prior assumptions about the nature of learning, intelligence, and development. These assumptions include, for example, the numbers of layers, the number of units in each layer, the extent of connectivity between layers, initial connection weights on units, initial activation values on units, the rate at which the network is permitted to learn, the way in which they are trained, and so on. These all affect the results, and are manipulated in order to nudge the network into behaving the way we want and producing the outcomes we then claim to be the miraculous properties of the network itself.

It is, perhaps, hardly surprising that some authors have criticized the haphazard and idiosyncratic nature of this approach. But this pragmatism leads to the second problem, which is that, when all is said and done, we remain remarkably naive about what is really being "learned" or "developed" in connectionist nets; how, exactly, it happens; and what is the form of the final representation. For example, Jeffrey Elman and his colleagues tell us that "knowledge ultimately refers to a specific pattern of synaptic connections in the brain," and that "knowledge is the representations that support behaviour." But what is the nature of these patterns and representations? What information, exactly, is represented? Remarkably, although much effort has gone into getting networks to work in a practical sense, there has been little analysis of the nature of the knowledge that results. If this is intelligence, we still know little about its real nature, except that it involves connection weights.

It has also been pointed out that connectionist modeling, so far, has been confined to "toy" problems, using highly tailored processing constraints. Most efforts have consisted of attempts to simulate well-known but relatively simple and isolated cognitive phenomena in equally simple networks (like the formation of prototype patterns described above). The prime motivation seems to have been to show that such networks can in principle do what the cognitive system does (with the implication that the latter must do it the same way).

Unfortunately, these simple networks rapidly break down when attempts are made to scale up to higher cognitive processes such as planning, reasoning, and theory-building. Furthermore, their neural and biological plausibility has been exaggerated. Consequently, many connectionists now see their models as nothing more than general schemata for what goes on in the brain. As one of the leaders of the field, Geoffrey Hinton of Carnegie-Mellon University, put it in an article in 1998: "I am disappointed that we still haven't got a clue what learning algorithms the brain uses."

Finally, connectionist models share a fundamental problem (already hinted at above) with symbolic models: they cannot act on the world. They can read it (or rather have it read to them), but they cannot change it. This sensation of action is a crucial aspect of intelligence and of what we know of how the brain works. Connectionist networks, too, remain passive, senseless "slaves" to the investigator's needs. As a model of human intelligence, they are surely a despot's dream!

It may seem somewhat extreme to claim that the two main branches of inquiry specifically attempting to describe the cognitive bases of intelligence have made little real headway, but many of their more honest proponents have come around to admitting that. I believe this limited progress is due to erroneous presuppositions about the functional requirements of the brain, and of the human variety of intelligence in particular. It is time, therefore, to turn to other models that try to overcome all these problems. Before we do, however, we need to spend some time examining the problem which I have several times hinted at as a stumbling block in all the models of intelligence examined so far—namely, the failure adequately to consider what intelligence is for, the structure of the environment, and its implications for cognition. This is what the next two chapters are about.

BIBLIOGRAPHY

Anderson, J. and E. Rosenfeld, eds. 1998. *Talking Nets: An Oral History of Neural Networks*. Cambridge, Mass.: MIT Press. A collection of candid reminiscences on the development and current status of connectionism from leaders in the field, including the article by Geoffrey Hinton.
Elman, J., E. Bates, A. Karmiloff-Smith, M. Johnson, D. Parisi, and K. Plunkett. 1997. *Rethinking Innateness: A Connectionist Perspective on Development*. Cambridge, Mass.: MIT Press. A detailed account of the modern connectionist view and an attempt to apply it to cognitive development.
Hanson, S. J. and D. J. Burr. 1990. "What Connectionist Models Learn: Learning and Representation in Connectionist Networks." *Behavioral and Brain Sciences* 13: 471–518. A review of principles and problems of connectionist models with wide-ranging peer commentary.
Harnad, S. 1990. "The Symbol-Grounding Problem." *Physica D* 52: 335–46. A well-known and concise review of one of the major problems of symbolic-computational models.
Newell, A. 1980. "Physical Symbol Systems." *Cognitive Sciences* 4: 135–83. A classic statement on cognition and intelligence as computations involving mental symbols.
Pinker, S. 1997. *How the Mind Works*. New York: Norton. A popular rendering of the evolutionary-computational view of intelligence.

5

Intelligent Systems

Herbert Spencer would have been delighted, no doubt, to find his *Law of Intelligence* (1855)—a correspondence between the "inner" and the "outer" orders—being so ably promulgated by the current clutch of evolutionary psychologists. "Our minds are designed to generate behaviour that would have been adaptive, on average, in our ancestral environment," says Steven Pinker. "Such brain mechanisms . . . that match the physical changes of the world are what we know variously and collectively as rationality or intelligence," says Henry Plotkin. Intelligence, or cognition generally, can be understood only as an evolved adaptation, say a host of other contemporary texts.

Indeed, the combination of Darwin's theory of natural selection and advances in genetics in the course of the twentieth century turns such stories into compelling ones. The genes, through the kind of mutations mentioned in chapter 4, and the recombinations that occur in the creation of eggs and sperm and at conception, can present considerable genetic variation. And some of that variation may be overtly expressed in variation in a character, or in some component of it. Some variants may be better adapted to aspects of the current environment than others. These favored variants will thus be selected in the sense that they are more likely to survive and leave surviving offspring. The gene or genes underlying the most "fit" variant(s) will thus increase in frequency across the generations (until every individual will possess it, or them). In this way, small advantages are accumulated and assembled into a well-adapted character, the information for which is transmitted in the genes across the generations. The origins of eyes, hands, feet, and every aspect of the mind are now explained by such accumulation of small advantages through natural selection.

Perhaps it is scarcely surprising that this compelling picture has been applied enthusiastically to our intelligence; or that intelligence, in turn, has been reduced to a standard adaptation in which natural selection has, in effect, constituted a set of genes (the genotype) which in turn determine the proteins, which determine the character (or phenotype):

natural selection $\rightarrow$ genotype $\rightarrow$ proteins $\rightarrow$ phenotype.

This, in turn, encourages the idea of the genes as a kind of recipe for intelligence. This is why Pinker speaks of "the genetic recipes that build the mind," and of the way that "our mental organs owe their basic design to our genetic program," and why it is now commonplace to think of the basic form of human intelligence as being "in the genes." As we saw in chapter 3, even authors who are seeking to revise the notion of innateness, and say they favor an interactive approach to the development of intelligence, cannot relinquish their belief in the genes as fundamental agents of developmental control. In such beliefs, then, all the complexities and specific regulatory roles of intelligence become collapsed back into a single framework. A mosaic of genetic information on the inside manifests itself as a mosaic of corresponding characters on the outside, with essentially the one, universal kind of relationship between them. This chimes in perfectly with the popular view held by the general public that our intelligence is essentially a product of our genes, with the environment attenuating or promoting their action but not altering the courses or end points of development. Thus the fatalism, already so much part of the social scene as well as psychological theory, becomes perpetuated.

In this chapter, we shall see that such a view, applied to all characters at all times, is highly simplistic and needs to be challenged. What do we mean by adaptation, for example? Hens' feet are adapted for scratching and perching. Ducks' feet are adapted for paddling in water. Eagles' beaks are adapted for tearing flesh. Finches' beaks are adapted for cracking seeds. Is intelligence an adaptation in this sense? Many, if not most, of the contemporary evolutionary psychologists seem to think so.

Also, in this chapter (and the next) we shall see why such a concept is a parody of any intelligent system. It is ironic that Darwin, in his preface to *The Origin of Species*, suspected that there were circumstances in which some system of adaptation other than natural selection of genetic variants might be needed. In this chapter we get a glimpse of what these other circumstances might be, and what they suggest about the nature of intelligence and of its regulations relative to those of other systems of adaptation. We will see that, instead of a "two-layer" relationship between genes and characters—genes on the inside, characters on the outside—that are vaguely affected by experience, complex organisms have evolved what have been described as a nested hierarchy of regulations. We will see how this became a necessary arrangement in increasingly unpredictable, changeable conditions, requiring adaptability as well as adaptedness. We will also see how such circumstances demand the origins of complex characters in processes other than the reading of recipes in genes and passive submission to the environment, but in a whole ensemble of interactive factors using and regulating gene products as developmental resources. Ultimately, these have produced an intelligence that anticipates, intervenes in, and positively fashions its own experience (even to the extent of changing its own genes, or their effects).

As mentioned at the end of the previous chapter, this approach will require rather more examination of the structure of the environment than is common in evolutionary psychological accounts, which focus on fixed adaptations by natural selection to relatively permanent aspects of the environment. The basic argument is this: all aspects of life are ways of dealing with incessant change. Without such change, life would be reduced to a conservative function of mere elimination of variants from static norms, with none of the diversity, the evolutionary complexity, and forms of intelligence that amaze us today. Indeed, it is thought that mechanisms for the production of genetic variation, such as the reassortments of genetic material that take place in the production of eggs and sperm, and sexual reproduction itself, evolved as adaptations for dealing with uncertain futures. These variants, although possibly redundant in the short term, can be precious resources for dealing with uncertain futures and new selection pressures.

In other words, adaptation by natural selection of variants is itself a mechanism for tracking environmental change. But tracking change by natural selection is essentially a trans-generational regulation: it cannot change what form an individual has in today's generation, but only what forms survive, and with what frequency, in tomorrow's. For these reasons, the mechanism can deal only with conditions that change very gradually and slowly (relative to the time it takes to reproduce the next generation). Natural selection as an adaptive mechanism can do little about changes that occur faster than that, such as any significant changes across the generation from parents to offspring, or any life-threatening changes within the lifetime of an individual. Dealing with these requires other, intragenerational regulations, in which lie, I argue, the seeds of all intelligence.

In other words, if we are to get a real purchase on the understanding of intelligence, it is essential that we recognize different systems of adaptation according to the qualitatively different kinds of conditions with which animals have had to cope as they have evolved, moving from relatively static, stable environments into increasingly nonstatic, nonstable ones. At the end of that evolutionary trail (which will become clear in chapter 7) is a whole new intelligent system that cannot be described purely in terms of biological laws; rather, a distinct layer of psychological laws becomes necessary. This realization, in turn, offers a glimpse of human intelligence, not as passive adapter to conditions "as given," but as a function able to anticipate and even make change. The purpose of this chapter is to distinguish and briefly describe those systems, and explain why they exist and how they have become functionally integrated.

Limitations of Genetic Instructions

The idea of genes as codes or instructions for characters is beguiling, and soon grabbed the attention of early geneticists. In the early years of the twentieth century, indeed, it was believed that each gene specified a single character, on a one-to-one basis. This soon emerged into the classic doctrine that different parts of the genetic material created the mature tissues and organs, so that there were genes for legs, for eyes, for brains, and so on.

It was eventually realized, however, that very many genes, perhaps hundreds, are involved in the development of complex characters. They are therefore called polygenic characters. It was discovered that such characters may or may not vary genetically in a population; that their development is influenced by the environment; that the same genes can be associated with highly variable formation of the phenotype; or, conversely, that even variable genes can be associated with a single, standard, phenotypic form (such as numbers of vertebrae or teeth, and so on). This all suggested rather more complex kinds of relations between genes and characters than the early doctrine had allowed. Contrary to the idea of recipes or programs, such well-established facts are indicative of other regulations in the formation of complex characters.

A moment's reflection will suggest reasons why the formation of characters must entail more than direct instructions from genes to characters. One of these is the fact that all cells contain the same genes, yet we get a vast variety of cells and tissues from them, not to mention complex physiological traits and cognitive structures and abilities. This differentiation suggests some regulations "higher" than genes as simple codes for characters. Indeed, a distinction has now been made between structural genes, which encode for proteins used directly for bodily structures and functions, and regulatory genes, the products of which determine if and when the others are translated. We will discuss these in greater depth below.

Another reason for questioning a simple instructive model of gene action is a logistical one. For example, the unregulated transcription of all genes at once would result in a logjam of products. Some products need to be available before others, but not before they are needed. It has long been recognized that the involvement of genes needs to be intricately timed, probably on the basis of feedback and signals from many other factors, including products already assembled. Hence we have the metaphor of a genetic program, an orchestration of gene action, as if it were in the genes themselves.

The genetic mode of adaptation would be unreliable in another sense: it assumes a reliability and regularity of internal and external conditions, so that genes can be mechanically translated and their products smoothly assembled. Such conditions may prevail for an egg dumped in an idealistic environmental soup. But for nearly all organisms, such conditions are extremely rare. More usual is a continual buffeting of physical bumps and shocks, of extreme swings in temperature, humidity, oxygen supply, and uneven provision of nutrients. Such buffeting would constantly threaten a process of direct translation from genes to characters, even for characters such as eyes and wings that are adaptive to relatively continuous aspects of environment.

This consideration applies to the construction of any complex structure. Consider a team of engineers erecting a crucial electricity pylon on a hillside in a gale, with heavy rain or snow. A stubborn commitment to a rigid series of step-by-step instructions would probably turn out to be fruitless, a waste of time and resources, and possibly of life itself. Instead, a flexible combination of steps (and ad hoc invention of new steps) responsive to current conditions and their ongoing change is required. This appears to be the kind of scenario described in recent molecular genetics of development, even in quite simple organisms, and it has transformed our understanding of the nature of gene involvement in development. In the rest of this chapter, we will look at two kinds of such interactions before turning, in the next chapter, to the development of intelligence itself.

Genomic Regulations

The nature of such interactions has been well demonstrated in organisms such as soil nematodes and the fruit fly called *Drosophila*. In these, and many others, it has been shown how development consists of rather more than the direct expression of structural genes into cell components, cells, tissues, and overt characters. This, of course, would result only in an ever-growing, but more or less uniform, ball of cells, devoid of head-to-tail or top-to-bottom axes for future development. Instead, the process consists of other regulations, which use the genes as developmental resources, not fundamental instructions or recipes.

For example, the earliest differentiation of the top (dorsal) and underneath (ventral) regions, and of the body axis in the fruit fly— with the head at the front and the tail at the rear—is not brought about by structural genes (whose products are incorporated into body structures) simply switching themselves on to produce the relevant constituents. It requires precise timing and spatial ordering of events. This even involves signals from outside the cell before the egg is laid. For example, early in egg formation in the fruit fly, future egg cells receive a chemical signal from surrounding cells in the mother's folli- cle, where the egg is produced. This chemical signal binds to proteins in the egg cell membranes, telling them, through a complex signaling pathway, whether they are to form dorsal or ventral regions of the egg.

The first differentiation of body form also involves the products of various maternal genes left in the egg before laying. These help regu- late, through various intermediaries, what the structural genes will produce and when. For example, the chemical product of the mater- nal gene known as *bicoid* appears to be critical in the formation of the embryonic head. It is present in the egg but, instead of being spread around evenly, it has its distribution constrained by the products of two other maternal genes, so that it is more concentrated in one part than another. This gradient of the product of the *bicoid* gene seems to be important in deciding which of the structural genes are recruited, and where and when, thus influencing the eventual differentiation of head and tail. Mutations deleting the *bicoid* gene and its product in the mother (and therefore its presence in the egg) results in the deletion of the entire head and thoracic structures in the embryo. They are replaced, though, not by a zone of undifferentiated cells, but by dupli- cated tail structures (so the fruit fly has no head and two tails). This suggests the mediation of still another raft of regulatory genes that influence what the structural genes actually produce, and where and when.

The form of regulation is also illustrated in the way that the differ- ent segments of a fruit fly embryo develop, with their appropriate internal organs. This involves (among many other things) the protein product of a gene called *Ultrabithorax* (*Ubx*), which varies across seg- ments in different concentrations at different times, with different consequences for the expression of a structural gene called *Antennape-dia* (*Antp*). As a result, the development of certain structures in some segments is repressed where, at that time, the concentration of the protein product of *Ubx* is initially strong, but not in others in which its concentration is weak. By the time distribution of the *Ubx* protein has become more even, the primordia of structures already forming are no longer repressed by *Ubx* protein. Thus, as Cambridge geneticists J. Castelli-Gair and M. Akam put it, the same structural genes can be related to different segment products in a way that "depends on the context in which they are expressed, and need not always have the same consequences in terms of 'segment identity.'"

That a given structural gene can be commandeered for different purposes is now a well-known phenomenon. For example, a variety of lectin—a protein known to have a defensive role against bacteria and other foreign substances in simple animals and plants—has been found to play important regulatory roles in development generally in flatworms and certain flies and cockroaches. Indeed, the origins of many human genes can be traced back over millions of years of evolution to analogues in flies and worms, although they may now have rather different functions. This flexibility under regulation is now known to extend even to the online restructuring of gene products themselves.

Structural genes are arranged on the chromosome in coding stretches known as exons interspersed by noncoding regions known as introns (figure 5.1). It seems that mechanisms exist for the rearrangement of the overall structure of the gene products by exon shuffling, so that different proteins, with different functions, are obtained from the same gene. As James Watson of Cold Spring Harbor Laboratory and his colleagues put it: "alternative splicing patterns . . . may represent a mechanism for organisms to adapt one basic protein structure (encoded by core exons) to different but related developmental purposes," whereby "the ability of the protein to interact with other cellular components could be changed as development progresses."

In addition, other developmental mechanisms have been identified that can modify the structure of the genes themselves; that is, to genetically engineer themselves within the individual according to current

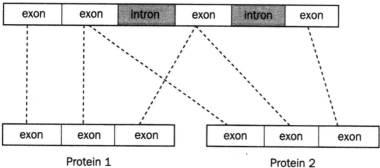

FIGURE 5.1. Production of different proteins from the same gene by exon shuffling induced by developmental regulations.

developmental needs. In his book *Phenotypes*, Canadian zoologist David Rollo says that, "Probably the greatest new development in our view of the genome is the diversity and action of transposable elements whose activities dice and splice the genome [and] modify the structure of genes themselves."

We have already seen how much of this regulation arises from outside the egg. The process of formation of different cell types and different tissues continues to be dependent upon external signals throughout development, including direct signals from the environment (such as light, temperature, spatial orientation, nutrient and other chemical gradients) and signals from other cells, close by and further away. Differentiation and development simply do not happen in the absence of such signals. Cells are unable to divide and differentiate unless they are embedded in a matrix with other cells from which they receive chemical signals. The whole process depends on complex controls from the outside, often referred to as social control.

Accordingly, it is increasingly recognized that structural genes are not independent, monarchic agents, churning out stereotyped trait components and commanding their arrangement in the developing organism. Rather, they are actually team players in complex webs of regulation, each being conditioned by others, and by their own products, in a dynamic, ongoing process. By the same token, we have to recognize that individuals receive far more than their genes from their parents. They also receive the whole complex machinery of a cell, comprising a host of active biochemicals, membranes and organelles, as well as products of their mother's genes: a whole developmental system. Development, instead of being seen as a passive product of genes and environment, has been moved to center stage as a set of determining interactions in its own right.

In other words, far from development being the passive product of structural genes and environmental fortune, we are beginning to see how its interactions and regulations are of the essence in the formation of a new individual. One indication of this is that up to 90 percent of genes may be regulatory in function (so only 10 percent of them are structural) and up to 90 percent of the structure of a typical gene has regulatory function, being responsive to regulatory signals from elsewhere, with only 10 percent that codes for the proteins that appear in body form. This is also suggested when amounts of DNA—the chemical of the genes—within the cells of different species are compared. Mice have 6,000 times as much DNA as bacteria, which is certainly consistent with differences in complexity. But humans have no more DNA than mice. The clear implication is that the gradual emergence of behavioral complexity within the mammals was not achieved by accumulating better single genes, but by increased regulation of those already available.

Given this whole ensemble of influences in the self-organization of development, it is easy to see why the metaphors of genetic control, genetic recipes, and so on are quite misleading. It is better to think in terms of genes as resources to be used when required in the dynamic process of development. It is difficult to describe the formation of even the most basic aspects of body form as that of a module which is the result of a genetic recipe. Likewise, in a context in which the same genes can be used in various ways, and various genes can be used for similar ends, the idea of a one-to-one relationship between genetic variation and phenotypic variation needs to be strongly qualified.

Indeed, long before these recent advances in developmental molecular genetics, the experiments of British biologist C. H. Waddington in the 1940s demonstrated the persistent development of certain characters in fruit flies, even when considerable genetic variability had been introduced by cross-breeding. In some experiments, it proved almost impossible to change the form of the character no matter what genes were present, and this applied in a wide range of environments. Waddington referred to the process as canalization, and described how developmental interactions acted to buffer the formation of body parts against genetic and environmental variations. Only in extreme conditions, such as extreme temperature, did the system break down and the underlying genetic variation show itself. This has been confirmed in more recent studies. In wild fruit flies, for example, the development of the eye, consisting of arrays of numerous subcomponents, is remarkably uniform, despite considerable underlying genetic variation.

What appears to have happened in evolution in changeable environments is that structural genes have been incorporated into higher genomic regulations, which dictate when and where their products should be forthcoming (figure 5.2). Such nesting of functions in a regulatory hierarchy has better ensured the development of particularly crucial characters in the face of environmental perturbations. It has also, of course, considerably enhanced the usefulness and likely survivability of single genes. Such a developmental system is far more robust in the face of local perturbations than isolated genetic instructions could possibly be.

In many quarters, such findings are now calling into question the

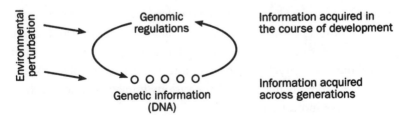

FIGURE 5.2. For the development of many crucial aspects of basic body form, under environmental perturbations, structural genes appear to have become nested in genomic regulations.

simple versions of neo-Darwinism as an accurate description of evolution and adaptation, so that some evolutionary biologists now speak of the post-Darwinian era. Instead of blind selectionism, which is both chancy and highly wasteful, there are creative powers in organisms, which theorists often refer to as emergent properties (of which more following). It is considered that these creative powers have themselves done much to determine the course of subsequent evolution. As we saw above, such creativity constitutes the seeds of intelligence, the opposite of passive dumbness.

Yet these points, so far, refer only to the problem of coping with environmental bumps and shocks that might threaten the development of basic body structures. As I hope to have shown, the problem is overcome by genomic interactions turning development itself into the coping mechanism. As we shall see when we turn to still more changeable environments, such a system of nested regulations becomes much deeper. Before doing so, however, it is worth considering other theoretical responses to the kind of intelligence just depicted.

Dynamic Systems Theory

The opposition to reductionism, like attempts to reduce the form of intelligence to a recipe in genes, has commonly consisted of appeals to organized structures and systems, involving ideas such as self-regulation and emergent properties, rather than independent determining factors. In biology, for example, Austrian L. von Bertalanffy's efforts to develop a "general theory of systems" based on a structured "organicism" helped to establish a broader stream of similar ideas. The wisdom of such a view has been urged on all of us in contemporary ecological disasters in which, for example, "smart" agrochemicals became widely used for pest destruction, crop fertilization, and so on, until it turned out that a much deeper set of interactions was involved. Much the same is now being realized for various human diseases, such as heart disease and cancer.

The most basic idea of self-organizing systems is that, instead of a single determining agent, there are several interacting forces, which can settle into a preferred state, known as an attractor. Among the myriad such systems described to date have been cloud formation, heart rhythms, electroencephalograms (EEGs), animal (and human) walking, and so on, as well as the genomic system just described. Each of these exhibits one or more steady states (the attractors). Although organized, and often developing in an orderly and highly adaptive manner, such systems have no controlling codes, recipes, programs, or blueprints, only internal interactions and their reactivity to external conditions. By analogy, development in biological and cognitive systems can be viewed as the transition from one attractor state to another as new conditions arise or are revealed by the individual's increasing action on the world.

MIT psychologist Esther Thelen and Linda Smith of Indiana University have made great efforts to describe human development in these terms, and offer samples of developmental data for interpretation in dynamic terms. One example is the case of infant kicking. When placed on their backs, infants perform rhythmic kicking movements, alternating between legs, which increase in frequency in the first six months of life. This basic, stable set of coordinations can be seen as an attractor state. But it is subject to increasing pressure from changes internal and external to the system. After about eight months, with increases in muscle and limb size, the coupling of movements in hip, knee, and ankle decreases dramatically, permitting more independent joint actions and thus new coordinations.

Such transitions indicate how new patterns of coordination can emerge from disturbances of an existing attractor state. Thelen and Smith argue that the development of limb coordinations from newborn kicking, through stepping, standing, walking, skipping, hopping, jumping, and so on, consists of a succession of such attractor states, each emerging as a result of changes (for example, in weight or size) in the limbs and joints and in their interface with the physical world.

Although such descriptions are intriguing, they have so far been much clearer, and thus more convincing, for physical systems (such as motor coordination) than for cognitive systems (such as intelligence). Of course, without clear description of the interactions in question, and the way in which they actually foster emergent intelligent properties, there is a danger, once again, of implicating mysterious, nonmaterial forces. We will look at one model of such interactions in chapters 6 and 7, after we have discussed other levels of regulation in development and adaptation.

Epigenetic Regulations

The hierarchical system of genomic regulations and the canalized development of necessary characters, described briefly here, have sometimes been referred to as epigenesis (meaning outside or beyond the genes). However, another kind of epigenesis has attracted just as much attention over the past few decades. To understand the nature of this system, we have to turn, once again, to consider environmental change. Although the genomic system buffers the developing embryo against local environmental bumps and shocks, the end state remains a predictable one, corresponding with reasonably predictable long-term environmental conditions.

Now consider what would happen if the environment itself, or a part of it important for survival, changed in important ways in the period between the parents' development and the offspring's development. For example, water may have dried up, foliage may have changed in color (ruining established camouflage), or the presence or absence of certain prey or predators may have changed. Such events are not uncommon in complex environments. In such circumstances, the canalized development of a specific end state is no longer adaptive.

Yet again, a solution to this problem has evolved. Instead of a specific end point, development of certain characters remains flexible until local conditions have been ascertained. Only then is an appropriate developmental pathway internally selected from a range of options. This is known as developmental plasticity, and the system is known as divergent epigenesis, in contrast with the convergent or canalized epigenesis already described. A dazzling variety of cases of developmental plasticity have been described in the scientific literature, but here we will confine ourselves to one or two simple examples.

In one sense, the phenomenon is shown in the diverse range of tissues of the body. Indeed, the first challenge to direct genetic determination of body form came when German biologist Hans Driesch, around 1890, separated the first two cells of a sea-urchin's egg and obtained not two half-urchins, but two completely formed organisms. This proved that each cell is capable of developing into any part of the body and thus contains a full complement of genes. It is therefore wrong to see development as the direct manifestation of a genetic mosaic within. Such pluripotency of cells, as it is known, and its dependence in signals from outside the cell, has been shown in many other ways. For example, cells removed from one part of the embryo, which might normally become the brain, and placed in another part, such as the skin, will duly develop an identity appropriate to their place. As Gilbert Gottlieb of the University of North Carolina whimsically explained: "We are sitting with parts of our body which could have been used for thinking"!

Divergent epigenesis has also been shown to determine individual differences in many physical and behavioral characters; for example, alternative wing patterns in moths and butterflies, fully winged versus flightless forms in many insects, variable defensive structures, variable feeding structures, and many others. It is important to stress that this is adaptive variation to local conditions and not just normal development attenuated by nonoptimal genes or environments. It is independent of genetic variation to a very large extent. For example, genetic variation in the terrestrial slug *Deroceras laeve* has been shown by Rollo and colleagues to be very low, yet the slug covers an enormous geographic range, and corresponding differentiation of form, thanks to developmental plasticity. So different are some varieties in some organisms, indeed, that they were once thought to be different species.

Sometimes the developmental plasticity is expressed as a quantitative gradient. For example, tails in mice are hairless and act as a heat radiator, the output of which can vary with tail length. Tail length has been shown to vary with the temperature of rearing. Similarly, the size of jaws in certain fishes in African lakes has been shown to vary according to the size of prey prevalent during development. In other cases, distinct phenotypic classes emerge according to conditions of development. Jean Piaget showed how the shell form in a species of snail can vary radically in structure—either a flat form or a spiked form, with different degrees of resistance to agitation—according to whether they developed in the calm waters of a pond or the choppy waters on the margins of a lake. A species of barnacle (*Chthamalus anisopoma*), studied by New Zealand biologist C. M. Lively, reacts to the presence of predatory snails in its environment by developing a bent form that is more resistant to predation than the more typical flat form. Oak caterpillars born in spring mimic the catkins (flowers) on which they feed, whereas those born in summer, and feeding on leaves, mimic twigs.

Again, none of these end states could have been precoded by natural selection in genes alone. Ecologist Richard Levins of Columbia University suggested the operation of a kind of developmental switch mechanism, triggered by an environmental cue: a wave of agitation in the case of the water snails, for example, or something excreted by predatory crabs, in the case of the barnacles, or by the concentration of tannin in leaves relative to catkins in the case of the caterpillars. Such environmental switches are most spectacularly seen at work in final sex determination in crocodiles and turtles, and caste determination in ants and bees, which is remarkably indeterminate until the constitution of the local population has been signaled.

Developmental plasticity has been well established in the brain. For example, genomic regulations seem to determine that cortical neurons form different computational types in the different layers of the cerebral cortex. But the response properties of local circuits, and the more gross differentiation of the cortex into specialized areas, seems to be determined by actual experience. In one set of studies by Oxford neurophysiologists Colin Blakemore and R. van Sluyters, cats reared in conditions in which they were exposed to lines of only one orientation developed a preponderance of nerve cells in the visual regions of the brain that were tuned to that particular orientation (as opposed to the full range normally developed).

It now seems clear that the differentiation of the cerebral cortex into functionally relevant areas arises not from a "protomap" laid down in genetic or genomic regulations, but as a result of specific forms of stimulation from the outside world. This has been demonstrated spectacularly in experiments by MIT neuroscientist Mriganka

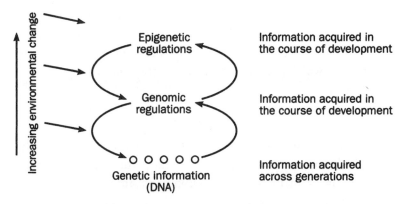

FIGURE 5.3. Nested hierarchy in epigenetic regulations.

Sur and his colleagues. They rerouted visual connections from the eye in newborn ferrets away from their usual destination in the visual cortex to what usually develops as the auditory cortex. The area that has long been recognized as the auditory cortex developed in these animals into a visual cortex instead. What is normally auditory cortex came, in these animals, to process and interpret visual signals. Similarly, it has been shown that plugs of cortex transplanted from, say, visual to somatosensory areas, develop connections characteristic of their new location, rather than those of their origins. Finally, it has been shown how the functions of one area that is surgically removed can be taken over by another. All this suggests again that, rather than being predetermined by a genetic code for cortical functions, these functions emerge from other regulations operating at critical times in the course of development.

In other words, we see again how adaptation has come to reside in broad developmental mechanisms, which, of course, include genes as resources but are not determined by them in any simple, direct manner. It is important to stress how these interactions have added further viability to those preceding them evolutionarily. By becoming embedded in a higher level of regulations, genomic (and genetic) regulations have been opened up and further extended (figure 5.3): they have become the subjects of, and resources for, organization, rather than its architects. Immediate reactions in the brain are first supported by local epigenetic regulations, which only later start switching genes on and off. It is an arrangement that evolved for dealing with environmental change with which genetic selection alone could not cope.

Lifelong Plasticity

These examples are cases of development which stop when the adult state has been reached. Yet environmental perturbation and change can continue, in many complex environments, throughout life. There are many cases of lifelong phenotypic plasticity. The term *cyclomorphosis* refers to cyclical changes, back and forth, in characters due to seasonal changes (for example, in temperature), or to the presence of predators or some chemical signals produced by predators. For example, the water flea *Daphnia* can change quite drastically to a more defensive form, depending on whether or not its common predator, the midge larva, is present in the water. Striking changes from flightless to winged forms in aphids and locusts are thought to be responses to overcrowding. Even more striking are the gross phenotypic changes that can occur in certain coral-reef fish. As described by Gottlieb (citing studies by D. Y. Shapiro):

> These fish live in spatially well-defined social groups in which there are many females and few males. When a male dies or is otherwise removed from the group, one of the females initiates a sex reversal over a period of about two days, in which she develops the coloration, behaviour, and gonadal physiology and anatomy of a fully functioning male.

Other examples of lifelong epigenesis, which are frequently taken for granted, are the myriad physiological characters that maintain the internal environment, respond to ingestion and digestion of food, respond to infection and disease, and so on, all of which have to do with rapid or sudden environmental changes that can occur throughout life. In each of these, the new level of regulation—a physiological function—is determining, rather than being determined by, epigenetic regulations (which in turn are determining genomic regulations and the recruitment of structural genes in an ongoing, regulated manner). The most striking example of lifelong plasticity is closer to the main concern of this book: behavior, and the cognitive regulations that govern it. Behavior emerged very early in animal evolution as a means for dealing with environmental change. At first, behavior consisted of simple reflexes for moving to a more favorable location, responding to temperature, light, and chemical gradients, and was coordinated through a simple nerve net. Eventually, it became more differentiated, requiring more complex coordinations, for more exacting purposes, in more complex, changeable environments. These raise a whole new layer of issues which we will start to explore in the next chapter.

BIBLIOGRAPHY

Corning, P. A. 1996. "The Cooperative Gene: On the Role of Synergy in Evolution." *Evolution Theory* 11: 183–207. On the limitations of independent genes and the dependence of complex systems on multilevel interactions.

Depew, D. J. and B. H. Weber. 1995. *Darwinism Evolving*. Cambridge, Mass.: MIT Press. Detailed review of cracks in the Darwinian monolith and emergence of "post-Darwinian" views of dealing with a changeable world, including the importance of hierarchical structures in many of them.

Goodwin, B. 1994. *How the Leopard Changed Its Spots*. London: Weidenfeld and Nicolson. Critique of Darwinism as a sufficient theory of the origins of form and its diversity, and powerful advocacy of dynamic, self-organizing principles in the development of living structures.

Gottlieb, G. 1991. "Experimental Canalization of Behavioral Development: Theory." *Developmental Psychology* 27: 4–13. A critical review of concepts of epigenetics as applied to human psychological development.

Lawrence, P. A. 1992. *The Making of a Fly: The Genetics of Animal Design*. Malden, Mass.: Blackwell. Fascinating review of genomic regulations in the development of body parts.

Oyama, S. 2000 (2d ed.). *The Ontogeny of Information*. New York: Cambridge University Press. Update of a classic critique of genetic determinism in development, and appeal to dynamic, self-organizing systems as the only guide to understanding.

Piaget, J. 1980. *Adaptation and Intelligence*. Chicago: University of Chicago Press. Piaget on epigenesis as a metaphor for the development of human intelligence.

Purves, D. 1994. *Neural Activity and the Growth of the Brain*. New York: Cambridge University Press. A readable overview of developmental principles in the brain.

Rollo, D. C. 1994. *Phenotypes: Their Epigenetics, Ecology, and Evolution*. London: Chapman and Hall. A thorough review of recent advances in molecular developmental genetics and how they transform traditional conceptions of how genes work.

Stearns, S. C. 1989. "The Evolutionary Significance of Phenotypic Plasticity." *BioScience* 39: 436–47. A readable review of developmental plasticity in plants and animals followed by peer commentaries, some considering implications for behavioral development.

Sur, M. 1993. "Cortical Specification: Microcircuits, Perceptual Identity, and an Overall Perspective." *Perspectives on Developmental Neurology* 1: 109–113. Review of research indicating environment-dependent epigenesis in brain.

Thelen, E. and L. B. Smith. 1994. *A Dynamic Systems Approach to the Development of Cognition and Action*. Cognitive Psychology Series. Cambridge, Mass.: MIT Press. Dynamic systems theory applied to human development.

Watson, J. D., S. H. Hopkins, J. W. Roberts, J. A. Steitz, and A. M. Weiner. 1990 (4th ed.). *Molecular Biology of the Gene*. Menlo Park, Calif.: Benjamin/Cummings. A comprehensive account of advances in understanding the structure and regulation of genes.

Weiner, J. 1999. *Time, Love, Memory: A Great Biologist and His Quest for the Origins of Behavior*. New York: Knopf.

6

Constructive Intelligence

In this chapter I want to extend the idea of a hierarchy of regulations, which I developed in the previous chapter, to cognitive regulations. Epigenesis has become a buzzword in psychology since the late 1970s, along with the idea that development of psychological characters is explained in the same way as the characters described under epigenetic and genomic regulations in the previous chapter. Although it is an obvious advance on the "summation of genetic and environmental charges" model, described in chapter 3, it is still an inadequate characterization of the regulations underlying human intelligence. In this chapter (and the next), we shall see why the epigenetic model is not the end of the story, and suggest what those additional regulations might be.

In the previous chapter, we saw that there are lifelong adaptable regulations (for example, physiological processes) which incorporate and extend epigenetic regulations, determining their activities rather than being determined by them. In this chapter, we shall see how cognitive regulations evolved as just such a system of lifelong adaptability, but for the regulation not of physiological functions or simple cyclical changes of form and behavior, but of far more adaptable forms of behavior. As before, such regulations have opened up the previously evolved regulations to allow an entirely new level of adaptability to new kinds of environmental challenge. Understanding this, however, requires further examination of the environment of cognition itself. In doing so, we come to realize that genomic and epigenetic regulations alone will not do for many conditions of life.

The Environment of Cognition

The common failure to examine the environment of cognition has created continual puzzlement about the nature and origins of intelligence. In my view, this is one reason for the tendency among psychologists to collapse everything back into a simple genetically determined story. We saw this in chapter 3, in relation to psychologists' vagueness about the environment of IQ development. Sometimes the puzzlement has been about why certain experiences (or lack of them) can seriously impair the development of the nervous systems of some animals but not others. For example, the American neurobiologist W. T. Greenough noted how rearing mammals in deprived environments is devastating to their cerebral development, yet has little effect on amphibians such as frogs. He therefore suggested that higher mammals and birds appear to be special cases in their prerequisites for normal brain development, and asks why and how, in the evolution of birds and mammals, experiential information sources have been so important to them, and apparently so unimportant to others.

Such questions are, of course, on a par with the puzzlement that prevails about why humans need the intelligence they have, and why they have such big brains. Despite strongly favoring a particular evolutionary story, Steven Pinker tells us that "the apparent evolutionary uselessness of human intelligence is a central problem of psychology, biology, and the scientific worldview." All this seems to bespeak rather impoverished conceptions of both intelligence and the environment to which it is said to be an adaptation. The dominant conception has been that intelligence equals learning, which equals the acquisition of simple environmental associations, such as learning an association between a signal and food location.

With this preconception of what the experiential information of intelligence consists,the British neurobiologist David Oakley tried to find a relationship between learning ability in various species and their (grossly different) amounts of neocortex. But he found little association at all: small-brained species appear to be just as good at learning as are big-brained species. Other researchers have trained animals in conditioned-response tasks, such as where to find food in a maze, and found that the trained responses were retained even after large parts of the cortex were removed. Adding to the puzzlement is the observation by other researchers that learning of quite complex associations occurs in molluscs. Oakley concluded that the kind of learning commonly studied in laboratories tells us little about what advantages the neocortex confers, and why it emerged in the course of vertebrate evolution.

All this is leading to the realization that the problem lies not in apparently superfluous brain tissue, but in our conception of what the experiential information of everyday action actually consists of. A more fruitful conception can be found, I argue, only in the environment of cognitive intelligence itself; in particular, in a description of the ways in which it is more complex than the environment of epigenesis. In this argument I assume we can take it for granted that the evolution of living things has consisted, more than anything else, of an increased ability to inhabit more complex and changeable conditions, and that the evolution of brain and intelligence reflects this. But what do we mean by complex, and how do we describe it? As the American (formerly Hungarian) systems theorist J. von Neumann said in the 1960s, without such an appreciation of complexity we can have little meaningful conception of evolution. Nor, in my view, can we have a meaningful conception of intelligence.

Describing Environmental Complexity

One approach to environmental complexity has been to describe it in informal terms, such as the numbers of objects in an animal's cage, opportunities for play or exploration, and so on. But such descriptions are obviously limited. They may help to identify factors in individual development, but they don't explain how they work, nor in what ways the environment of apes may be more complex than that of rats or mice, say (nor, therefore, why they are more intelligent).

So what else can we mean by complexity? A broad way of looking at it would be to argue that all living things need their environments to be predictable for life to be possible at all: what organs, appendages, and processes should be developed and what behaviors planned? So complexity must be related to the relative ease or difficulty with which predictability can be found in a habitat. A more principled way of describing this would be to enumerate all the factors or variables in an animal's world, and the nature of the relations between them which it would be necessary to take account of, for predictability to be possible. Indeed, the description of a system in terms of the interactions between its components or variables has long been used as a natural measure of complexity by information theorists.

Let us look at this idea more closely. All predictability arises from associations between things that vary; for example, you can predict (to a certain degree) the likely weight of a person from his or her height. We call the dimensions variables and, because they vary together to some extent, we say that they co-vary, or exhibit a degree of covariation. It is because they co-vary that you can tell, to some extent, the values on one variable from values on the other. A correlation, as discussed in chapter 2, is a special kind of covariation which measures only linear relations (a fixed increment of height is always accompanied by a fixed increment in weight), whereas here we will be discussing all possible covariations.

Much predictability in nature may arise from simple covariations between two variables—for example, between a sound and the presence of a predator, or between a location and food. As mentioned above, much learning theory has been based on getting animals to become sensitive to such simple covariations in laboratory training boxes. In other situations, though, adequate predictability may be available only in much deeper covariations between variables. Consider a very simple case: daily (diurnal) or seasonal changes. Figure 6.1 shows how there may be no direct covariation at all between two variables (place and food); at first glance, there seems to be no way of predicting whether food is going to be in a particular place at all. Yet a more detailed analysis shows that there is covariation, and we can predict the presence of one from the other after all: it only appeared elusive because it is nested in, or conditioned by, the other variable, season.

A simpler way of putting this is to say that the covariation between food and place depends on the season. However, these are all ways of

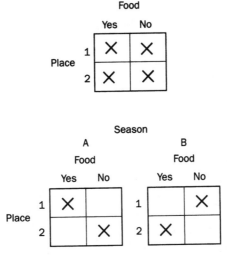

FIGURE 6.1. Predictability between two variables (food and place) is absent, unless its nestedness in another variable (season) is registered. Predictability arises from this interactive relation (crosses indicate co-occurrence of the variable values).

saying that there is an interaction among the three variables. This interaction is an important, indeed crucial, source of new information for predictability. The important point is that only by taking account of the interaction can the whole situation become predictable. As we know, the physiological systems of migratory birds and hibernating mammals have evolved to take advantage of such interactions. By being sensitive to cues for seasonal changes, such as length of daylight, they are predicting when food will (or will not) be present in a certain place, and respond accordingly. Such interactions can, of course, lie much deeper than this example suggests, involving many more variables, with many more values and nonlinearities, and this is what describes the environment's complexity.

In the example just given, the overall structure of interaction—the covariation between food and place, and its dependence on the season—remains constant over relatively long periods of time. This is the main reason it can be coped with, and appropriate behaviors regulated, by an equally constant physiological system. Indeed, behaviors that have evolved at that level, responsive to relatively continuous contingencies and persisting across generations, are often distinguished by the label "instincts."

But other conditions of life can be much more complex and changeable than this. Covariations and interactions between variables may themselves change frequently in the course of a single lifetime. Indeed, animals that behave usually change conditions themselves by their own activities. Predictability is still a prerequisite for survival, but it now has to be found in deeper interactions among, potentially, many more than the three variables shown in the example above. And because the whole structure of relations can change, any predictability has to be induced, as it were, on a continuously updated basis. This is a different kind of complexity, and it requires a different system for rendering predictability: a cognitive one.

Intelligence and Environmental Structure

This is what I will be referring to when I talk about environmental structure (a term that is often used, but usually imprecisely). Perhaps the very simplest example of learning by abstraction of deep covariation structure, in constantly novel form, is shown in the alternation task, which normal monkeys, as well as humans, readily learn. In this task, which they will never have experienced before, a reward, such as a peanut, is placed alternately under one and then the other of two inverted cups behind a screen, before the screen is removed. Again, there is a covariation between food and place, but this time it is one conditioned by the alternation imposed by the experimenter. To get the reward consistently, the animal has to induce (cognitively) this informational structure and, on each trial, predict its new location. Monkeys that have sustained injuries to their frontal lobes are completely incapable of performing this task, although their ability to learn a range of more direct stimulus-response associations remains unimpaired.

To illustrate a more complex case, consider a barn owl, which will, at any moment, be flying or sitting at some height off the ground, seeking cues to prey, which are not stereotyped because the prey itself will be moving, partly concealed, at some novel distance, angle, and so on. In studies of the barn owl's ability to predict the location of prey from sound, the Stanford neurobiologist E. I. Knudson has shown how prediction doesn't arise from a simple cue-response function. No such covariation exists in any useful sense in the typical relation between owl and prey. Rather, predictability is possible only because of the deeply structured interactions between a whole set of variables.

For example, there will be a relation between sound intensity at the two ears, such that the difference in intensity will co-vary with the direction of origin of the sound: the further the prey lies to left or right, the bigger the intensity difference will be. This covariation can be used to predict the location of prey. However, this covariation also depends on the distance of the prey from the bird. There is an interaction such that, the greater the distance, the lower the actual intensity difference at the ears is likely to be. It is also the case that this interaction itself depends on the frequency of the sound, so that there is a still deeper interaction. Finally, this whole interaction structure itself depends upon (interacts with) the developmental age of the bird, and the growing distance between the ears. Only by inducing an internal representation of this whole interactive structure can the location of the prey be predicted from the sound with an adequate degree of precision. Because this representing has to be done on a continually renewable basis, it explains why even the barn owl needs a cognitive system. When it starts to use it in this way, it starts to be intelligent.

Such intelligence is crucial in animals that alter the world by their own activities. Consider, as a hypothetical example, a monkey that feeds on fruit in the upper canopy of a rainforest (having discovered a covariation between elevation and food abundance). The monkey also registers the fact that, nearer the forest margins, the fruit lies lower down (the height-abundance covariation is conditioned by location). But those marginal locations, at certain times of the year, are also more frequently scanned by monkey eagles (the conditioning is itself conditioned by a covariation between predator presence and season). This whole covariation structure may change completely when the troop moves to another forest because of fruit depletion (the monkeys' own activity) or perhaps seasonal changes.

The problem becomes far more complex in animals that need to do more than predict just the location of objects (such as prey), but also many other possible properties as well. What is it, actually? Can I eat it? Is it dangerous? Will it be heavy? It looks small from here, but how big is it close up? Can I use it as my new nest? Can I throw it? All such questions are hugely complicated by the fact that objects are usually experienced in dynamic, constantly novel forms; just establishing a stable image of it is a major cognitive task. We saw in chapter 4 how this dynamic reality is devastating to computational models of intelligence based on fixed If-Then cues. Making sense of such a world depends on inducing a representation, or dynamic model, of the covariation structure of objects and events as experienced in ever-changeable forms, rather than fixed cues.

Animals living in complex environments need to be able to abstract and represent conceptually such covariation structures because these are often the only consistent information available. I have called these kinds of information structures hyperstructures (although I will sometimes call them covariation structures, or nested or embedded covariations, and so on). By representing the outside world of objects and events in this language, form is captured and preserved despite constant transformation of its superficial appearance. Thus, each and every object may be experienced in fleeting and partial images that are constantly novel, but the covariation pattern within them remains the same, and is characteristic for a given object. Once abstracted, this informational structure serves as a kind of grammar from which all future images of objects and events can be created, even when, as is often the case, the input is sparse or fragmentary. A current representation of an object or event is thus a construction, a result of the interaction between the covariation pattern in current input, and that in the internal representation of the corresponding object or event, built up from previous experience.

The remarkable constructive power of such a system is indicated in experiments using point-light stimuli, which we saw in chapter 4. These are usually made by attaching ten or so bits of reflective material to various parts of the object, such as the joints of a person walking, or the corners and edges of an object as viewed in normal use, and then filming in a darkened room while a light is being shone on it, so that only the moving point lights can be seen. When the film is shown to other people, they very quickly recognize the source object, even though what is seen is an extremely skimpy version of the original.

In experiments with David Webster at the Open University (U.K.), we showed that adults and children continue to recognize, quite quickly, an object's form from stimuli consisting of as few as four points, despite the fact that the stimuli contain no overt features (of the kind needed by computational demons). We suspected that what enabled people to do this was the presence of deep covariation structures in the stimuli; for example, covariation between the movement of two points is conditioned by movement of a third point, and that interactive structure is conditioned by a fourth, and so on. We suspected that it was the match between this covariation structure and that in an internal representation of the same object (a person walking, say) that was leading to recognition.

Indeed, when we looked for such deep interactions between the movements of points in the stimuli, using mathematical techniques, we found them in abundance. More important, it turned out that the number of people recognizing the source object was significantly related to the complexity of those interactions in the stimulus (as just defined). Thus it seems that recognition of objects, even from stimuli as sparse as a few light points, can occur because long experience with the object has built up a cognitive representation of the complex covariation structure it exhibits. A sample of that structure in the stimulus activates the corresponding structure in representation. The detailed covariation structure therein is then used to predict all the missing parts in the current (very sparse) stimulus, and a full image (a person walking) is created online.

A Systems Definition of Intelligence

When animals began to evolve in more complex environments, predictability could not always be found in direct or simple associations. It had to be found, instead, in the interactive structures between associations and variables, some of them lying very deep. In my view, this is what a cognitive system is for. I think such a view takes us a little nearer to a useful definition of animal intelligence (although, for reasons explained in chapter 7, not yet all the way to human intelligence). Genetic regulations consisting of the passive generation of genetic variants, acted upon by natural selection, can match a character to stable or gradually changing conditions. Genomic regulations can buffer the development of crucial characters against genetic variation and environmental disturbances during development. Epigenetic regulations furnish a plasticity of development in the face of more rapid changes across successive generations, or to lifelong changes that are fairly shallow (for example, cyclical, diurnal, and seasonal changes). But none of these can furnish the capacity to deal with deep structural associations, which are the only sources of predictability in conditions that change rapidly throughout life.

I think this view also helps to define the terms in which interspecific differences in intelligence can be described quantitatively—that is, the depth of the interactive relations between variables that can be grasped. By this I mean the ability to grasp that simple or superficial associations (between two variables, say) may themselves be conditioned by other variables, and that this conditioning may be conditioned by a still deeper interaction with other variables, and so on. The need to capture and constantly update such information structure in ever deeper complexity may also explain the rapid evolution of brain size in mammals.

Although human intelligence operates at a completely different level, as we shall see in chapter 7, this kind of complexity within it is evident from numerous studies. For example, Ceci and Liker's study of betting at racetracks (described in chapter 2) showed that subjects used representations consisting of deep, often nonlinear, interactions involving up to eleven variables. It is easy to find myriad examples, from birds flying through trees, to people driving on busy motorways, in which constantly novel current images have to be rapidly interpreted from representations of informational structure built up from past experience. Adaptation depends neither on innate programs nor empirical copies, but on representational structures of increasing depth and sophistication developing on a lifelong basis.

It is important to stress that cognitive regulations have not eclipsed

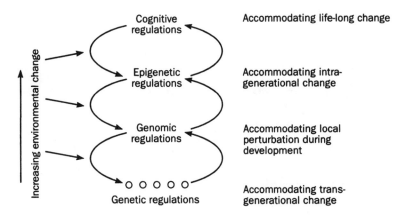

FIGURE 6.2. Nesting of genomic and epigenetic regulations in cognitive regulations.

epigenetic regulations. Instead, the latter have become nested within them and, in the course of evolution, opened up and extended to serve more demanding ends (figure 6.2).

In this arrangement, epigenetic regulation and plasticity, seen previously only in the developmental period and confined to a limited range of forms, can now operate throughout life in generating a far more adaptable intelligence. This occurs mainly in the central nervous system: lifelong epigenetic changes in cortical connections and neuronal response properties, operating under the guidance of cognitive regulations, are now well-established findings (of which more in chapter 8). Likewise, there is considerable evidence in the adult brains of mammals, not just of reconnections among existing circuits in the brain on a lifelong basis, but also of rapid formation of new connections, generated locally, on demand. Under cognitive regulations, the mature adult brain maintains plasticity that was previously confined to juveniles in the course of early development.

Modeling Cognitive Regulations

Some recent speculation about the nature of intelligence has certainly included a more healthy emphasis on complexity, change, representation, and the uncertain-futures problem. In his book *The Thinking Ape: Evolutionary Origins of Intelligence*, the British behavioral biologist Richard Byrne says that a suitable definition of intelligence includes "learning an unrestricted range of information; applying this knowledge in other, perhaps novel, situations; profiting from the skills of others; and thinking, reasoning or planning novel tactics." Mark Johnson of Birkbeck College (London) says that intelligence "refers to a sophisticated capacity for representation that enables an animal to be especially flexible in applying what it has learned to novel situations." Despite ultimately settling for an adaptationist stance, as mentioned earlier, Henry Plotkin says that "predictable unpredictability is the core concept for an understanding of why intelligence . . . evolved; it is the condition of the world that instincts as adaptations cannot deal with."

The problem is that none of these appeals offer a principled account of the nature of this flexibility or sophisticated representation, or the source of predictability in complex environments, or how we get a creative and flexible intelligence from it. However, many psychologists have offered constructivist accounts of intelligence to rival those of the computationalists and associationists described in chapter 4. These accounts not only reflect many of the properties just described, but extensions of them have attempted to show how real intelligence operates in real contexts. Because the best-known constructivist model is that of Piaget, we will use it to illustrate the point, and then move on to how other theorists have brought in the role of context. We will look at gaps in these theories and, in the next chapter, we will examine them more closely by considering human social intelligence.

Piaget's Theory

For Jean Piaget, human intelligence cannot be described either as the operation of a priori structures (such as innate modules or constraints), nor as shallow associationist copies of the world. He said that intelligence can be likened to a continuous process of adaptation that characterizes so much of life in general. But, unlike the maturation of preadapted physical processes of the body, adaptation at the mental level cannot be predetermined. The evidence for this is that, in the emergence of knowledge and reasoning powers in individuals, there is a continual construction of novelty. As he put it in his book *Epistemology of Human Sciences*: "The operatory structures of intelligence are not innate. . . . They are not preformed within the nervous system, neither are they in the physical world where they would only have to be discovered. They therefore testify a real construction."

The most basic ingredients of this intelligence, according to Piaget, are the deeper "coordinations" in the physical and social worlds, over and above the information in direct perception. These are what are revealed in everyday actions, and become represented in mental structures or operations. When these coordinations have been fully grasped, the representations are in equilibrium, and it is this equilibrium that Piaget saw as the prerequisite for logical and scientific thinking.

Take, for example, a ball of clay rolled out into successive shapes. According to Piaget, the actions of the subject on the object reveal not just isolated properties of the clay but the coordinations between them. The length and the thickness of the ball are not independent dimensions. There is a necessary connection between them: as one changes so does the other—that is, they vary together, or co-vary. In other words, they are coordinated. But this coordination is itself only part of a wider system of coordinations of which even a simple action like this consists: coordinations between the visual appearance and the motion of the ball, for example, and between these and the sense receptors in skin, muscles, and joints. Such coordinations are, of course, very similar to the dynamic associations we looked at in the previous section.

Of course, the shape of each ball of clay that we encounter (its length relative to thickness) will be novel. So if we want to do something with it (lengthen it, shorten it, or whatever) then we need to have a knowledge representation consisting of more than a fixed rule responding to a fixed cue, as each cue will almost certainly be novel. The system needs to have captured the more abstract coordination between the variables involved. This is what distinguishes an intelligent system from series of knee-jerk reflexes, as favored by computationalists. The really significant point is that, when these coordinations have been represented in the sensorimotor system, a set of new powers becomes available that vastly increase our predictive abilities (our intelligence) about the world. For example, there is now compensation between the variables. We can predict that reducing the length will increase the thickness proportionately, and vice versa. In addition, we can predict what effort would be required to return it to its original shape. The structures of intelligence thereby capture the dynamic reversibility between states.

In everyday thoughts and actions, such as digging, riding bicycles, lifting objects, and so on, in which the myriad permutations of sensations and motor actions are almost always novel, we take these powers for granted. But they would not be available without the representations of coordinations being constructed. This is why coordination is such an important idea in Piaget's theory.

In the first months and years of life, children acquire reversibility at a sensorimotor level (in action). Through continual reequilibration of coordinations, Piaget argues, children develop the classic representations of space, time, and causality. Thus the newborn, and young infant, confronts a world experienced as an incoherent stream of sensations. At this stage, not even a distinction between the infant's own body and the outside world is made, and the first few months consist of a rapid accumulation of sensorimotor structures or coordinations.

A good example is the concept of object permanence. In one famous demonstration of behavior before this stage, an infant of four months, say, is presented with an object. Just as reaching commences, however, the object is covered with an inverted cup. Thus concealed from sight, the object has no further existence as far as the infant is concerned, and reaching ceases immediately.

With the development of appropriate structures (intercoordinations), however, the infant appreciates that what is seen or not seen is a consequence of relations between objects in a coordinated system, rather than one set of fleeting stimuli replacing another. Reaching for the object continues, even though it is now hidden. This reflects a new operational structure that involves not just the idea that a body exists but also a whole system of spatial awareness from which a new range of predictions can be made on the basis of the spatial relations among objects. One striking consequence, according to Piaget, is that the

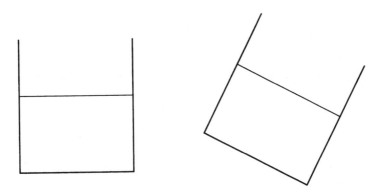

FIGURE 6.3. The failure to fully represent coordinations in an inclined beaker of water.

infant can now conceive of him or herself as an object, distinct from other objects in the outside world, a momentous development which he called a "Copernican revolution."

Although captured in some form in sensorimotor intelligence, none of these structures or sets of coordinations is yet available explicitly in thought, which is what Piaget meant by representation. This representation at the higher level does not appear until around seven years of age, according to Piaget. A classic illustration of this delay is the four-year-old who understands in action all the coordinations of forces involved in drinking from a beaker, yet when asked to draw a beaker of water tipped to one side, presents a picture that fails to conserve them (figure 6.3). Piaget said that to achieve this higher level of thought, the child has, in a sense, to do it all over again, in internalizing the coordinations of action from a sensorimotor form into new forms of representation.

Another illustration of this problem, according to Piaget, is the conservation of form across transformation of perceptual states. For example, a five-year-old may be shown two rows of beads, as in figure 6.4, and asked whether the rows have the same number of beads (adjustment is made until that agreement is reached). One of the rows is then extended, as shown in figure 6.4, and the question is repeated. Younger children fail to conserve by saying that the transformed row

The child agrees that
there is the same number
of beads in each row

One row is 'spread out' more and
the child is asked whether there is
still the same number in both

FIGURE 6.4. Test for conservation of quantity.

now has a greater number: the child seems to fail to coordinate the two states through the action of transformation, relying purely on perception of the independent dimensions.

Such coordination in representation (in the form of what Piaget called logico-mathematical structures) also governs the development of other concepts, such as classification (the coordination of subclasses into classes), seriation (the coordination of a series of objects in some order, such as increasing size), and number. These all entail the development of logical thought, not just in the sense of internal consistency but in the child's awareness of the necessity of its implications. These structures develop very slowly up to the age of seven or eight years. But even up to the age of eleven years or so, they are expressed in thought and action only with concrete objects and situations that are immediately present in experience.

Only after eleven years of age, when these separate logico-mathematical structures have themselves become fully intercoordinated, can mature reversibility in thought be found. Intelligence can then become detached from concrete instantiations. When this has happened, imagined or hypothetical situations can be entertained, and so take on the qualities of mature logical and scientific thinking, as in recognition of the need to control variables in scientific tests of relationships. However, it must be stressed that age has simply been a convenient descriptive marker and is not itself an actual criterion of developmental stage.

Criticisms of Piaget

Although Piaget's theory easily remains the most comprehensive theory of cognitive development and intelligence yet proposed, it has proved to be controversial. One of the major criticisms has concerned aspects of his methodology, and thus of his empirical base. Critics have argued that at least some of Piaget's findings—for example, the conservation task just described—are artifacts of asking children strange questions about strange things in strange situations. Modifications of the tasks, which introduce familiarity of task structure or context, transform children's performances, and indicate the development of conservational and other logic much earlier than found by Piaget.

For example, French neuroscientist Stanislas Dehaene, in his book *The Number Sense*, reviews evidence that appears to support the idea of a number concept in infants just a few days old. Such evidence relies on the looking-time methodology described briefly in chapter 3. Infants are habituated to pictures showing two objects (such as dots or toys) and, after their looking time has waned, they are then presented with a picture showing *three* objects. A much-recovered looking time has been reported. The infants are said to show surprise and thus the ability to distinguish between two and three. Other manipulations of the looking-time methodology have been taken to suggest that infants four to five months old can even do elementary addition!

There are several problems with this interpretation, however. One problem arises from doubts about infants' visual acuity in the first few weeks of life. Another arises from the difficulty of getting reliable results from infants at this age because of "fussing" (which is such that up to 50 percent of participants may have to be rejected). The most fundamental objection, though, is that, just because a discrimination between two objects and three objects is made, it doesn't necessarily mean that the infant has a number concept. It may be due simply to a general variation discrimination residing at a sensory or perceptual level. After all, variation discrimination is the evolved, specific property of every kind of sensory organ (organs of vision, hearing, touch, taste, and so on). We could equip a machine with the same kind of powers of discrimination without concluding that it has a number concept. As mentioned in chapter 3, Richard Bogartz and his colleagues concluded that all the results of the smart-infant studies can be interpreted in this more basic way.

This is why Piaget insisted that we check that the underlying logic of a number concept, and not just a general variation discrimination, is present before we conclude whether or not it exists. In his own theory, the logic involves having the power of reversibility between two competing states (two different spreads of a row of marbles, for example), and that the relationship exists of necessity. This is why he always asked children to justify their answers in tests like that for number conservation, and concluded affirmatively only if they expressed that necessity factor (for example, "you just spread them out more, so the number must stay the same"). In their book *Children Doing Mathematics*, Terezinha Nunes of London University and Peter Bryant of Oxford University stress how the understanding of such a logic of number is a vital prerequisite to the learning of mathematics in schools. Piaget said we must verify the presence of their underlying logic in the other concepts he studied before we declare that they are present.

In one sense, the development of intelligence in early childhood is so dazzling that it's easy to have some sympathy with theorists who appeal to hidden agents of design. But it is just as important to remember that the latter remain far more products of imagination than exact description. Dehaene, for example, reiterates the classic argument in suggesting that the infant brain "comes equipped with numerical detectors" laid down by a plan "that probably belongs to our genetic endowment." But he also says that this is only a sketch of a "tentative scenario." How an abstract concept such as number can be encoded in a linear strand of DNA is not explained. As we shall see in chapter 7, the crucial factor in the evolution and development of number was the unique degree of social cooperation and communication in humans, building on the more basic general ability for the detection of variation.

A possibly more justifiable criticism of Piaget's theory concerns his attempts to characterize the actual content and logical aspects of intelligence (such as number) in formal terms. What forms these take in the head is uncertain. Throughout his academic life, Piaget used various mathematical and logical theories (theory of groups, of propositional logic, of "morphisms," and so on) to attempt to describe these logical structures. Uncertainty about whether these are mere descriptive formalisms, metaphors, or exact models of the constitution of intellectual structures has been widespread. I suggested here that the content of these mental structures may actually consist of the deep covariation structures present in, and abstracted from, the world of experience. These would show all the reversibility and logical properties that Piaget demanded, but in a tractable form.

Some psychologists have criticized Piaget's theory as an overly individualistic one in which the role of social cooperation in logic formation is neglected. Such criticism may be unfair. Leslie Smith of the University of Lancaster has pointed out in a number of works that Piaget's thinking in this area has been grossly underestimated. It is also important to point out that Piaget continually responded with further modifications to his theory and new descriptions of logical structures. The most recent of these appeared for the first time in English translations as recently as the 1990s and appear radical enough to have been described as "Piaget's new theory."

Finally, an increasingly popular argument in recent years has been that the constructivism that Piaget proposed cannot be described purely in terms of internal adjustments to external structures. Rather, we need to take more account of the contexts in which thought develops, and which may be reflected in more context-specific mental structures. A number of inquiries (which we shall look at in a moment) appear to have demonstrated an important role of current context, including cultural context, as well as other specifics of past experience, as important means of shaping current logical activity.

Intelligence in Context

Piaget's theory, like all the theories of intelligence described so far, implies the existence of general, logical rules of thought, intrinsic to or developed in the cognitive system, as the main vehicles of intelligence. The problem is that the search by psychologists for universal laws of thought or logic that operate independently of context and content has proved to be consistently frustrating. Following one series of studies in the 1970s, British psychologists Peter Wason and Philip Johnson-Laird described how it was impossible to find a formal calculus that correctly modeled their subjects' inferences. Although shaking the most fundamental assumptions on which so much theory about intelligence has been built, the challenge has been greatly reinforced by other research over the past two decades. Here I can only give brief illustrations.

In one study by the American psychologist Stephen Ceci and his Brazilian colleague Antonio Roazzi, children were presented with geometric shapes moving on a computer screen, the movements dictated by an underlying complex rule imposed by the investigators. Children had to make predictions about positions of the shapes and, even after 750 trials, they were still predicting at just above chance. But performance improved dramatically when the shapes were substituted with model animals (bees, birds, and butterflies) and incorporated into a game in which they had to be caught in a net, although exactly the same underlying rules of movement remained. In this context, children were able to bring their background knowledge to bear on the problem, use it to help induce the underlying structure, and deduce accurate predictions from it.

A number of studies have shown similar effects with Piagetian tasks, such as the conservation task described above. British psychologist Paul Light and his colleagues had pairs of children racing to fill a grid with pasta shells, the initial equality of quantity being ensured by measuring into similar beakers at the outset. The children agreed to this equality. Then the investigator "noticed" that one of the beakers was chipped, and safety demanded that the contents be transferred to another, larger beaker (the only alternative available). The shells were duly poured from one to the other. After questioning, many more children agreed that the quantities were still the same than had done in the pretest using the standard procedure.

Such effects seem to be particularly strong among individuals steeped in knowledge in a particular domain. It is well known that young children, who obviously score less than adults on IQ tests, nevertheless perform much better than adults on tests involving a domain (such as computers or chess) in which they are particularly interested. Likewise, although high IQ is popularly thought of as reflecting the ability to think logically, there tends to be little relationship between IQ test performance and real-life logical performance. Various examples of this anomaly were presented in chapter 2.

Other studies have shown important effects of context and background knowledge on IQ test performance, especially the nonverbal forms (which, as we saw in chapter 2, are considered to be "culture-free" and thus particularly indicative of the ability for logical thought). For example, Usha Goswami of University College (London) has shown the effects of context and background knowledge on a wide range of analogical reasoning tasks. In another study, I took ten of the most difficult set of the progressive Raven's matrices (probably the most famous, and widely used, "culture-free" test) and devised a parallel set, including the same reasoning structures as the original items but couched in familiar contexts rather than abstract geometric figures. Eleven-year-olds performed much better with the new items than with the former, although they required exactly the same logic, and there was no relation between performances on the two kinds of tests.

Such findings simply confirm a general facilitation of familiar context and background knowledge, of course, without indicating how they work. This still remains a stumbling block in the area, as it does in theory of intelligence and cognition generally. As British psychologist George Butterworth has pointed out, simply to insist on putting thought into context is not enough to explain its nature and development. And, as Ceci and Roazzi have explained, although studies regularly show changes in performance as a function of changes in context, we have little idea of how this happens.

A number of theories have been proposed, however. One of these (proposed by Johnson-Laird and others) suggests that background knowledge allows the situation described in the premise to be assembled into a kind of mental model. This model can then be scanned for alternative predictions. Most of the research on mental models has involved the use of logical syllogisms containing certain premises entailing one or more conclusions that have to be inferred. This can best be seen by example.

All artists are beekeepers
And all beekeepers are chemists
Are all artists chemists?

In a number of studies, Johnson-Laird and his colleagues have shown how reasoning is affected not simply by the premises, but also by the ease or difficulty with which a mental model of the situation can be constructed from them. They thus suggest that the development of reasoning ability depends less on the acquisition of formal rules of logic than on procedures for constructing and manipulating such models.

Other theorists have suggested a process in which the structure of a given problem, with the assistance of contextual cues, maps on to knowledge structures derived from experience in the domain, these structures in turn generating predictions and solutions. For example, if the structure of a syllogism or other logical problem mimics that of a common social scenario or social convention, then the solution is found more readily. American psychologist Jerome Bruner has pointed out how many people err in the following problem:

All A's are B's
Are all B's A's?

But few fail when the same problem structure is couched in a familiar context:

All humans are mammals
Are all mammals humans?

The problem with these proposals is, again, that of the form they take in the mind or brain. And how, exactly, is context subsumed within it? Some individuals who appear to be able to complete the more abstract version of this syllogism above, for example, may not be doing it by virtue of being better abstract thinkers, but because they are used to solving problems of this type and are using a familiar context as a framework, by doing it covertly. These are formidable questions. But I have already hinted at the most important—indeed, the omnipresent—context of human intelligence, which is that of people's sociocultural history. This, and what it tells us about the nature of intelligence, is what we shall move on to in the next chapter.

A Creative Intelligence

Despite problems over these details, it seems important to emphasize again how these constructivist proposals counter those of the computationalist, the biological reductionist, and the IQ tester. As mentioned in chapter 4, one of the problems with the evolutionary, adaptationist view is that it depicts a passive intelligence, submissive to whatever problems the environment poses, responsive only with pre-structured or constrained routines. But the essential problem of a cognitive system is precisely one in which the whole complex structure of experience changes repeatedly, throughout life.

A constructivist system turns its bearer from a merely passive adapter to the environment into a much more active one. Piaget described the brilliance of this system even as it emerges in human children in the sensorimotor stage of development in the first year or so of life. The onset of pretend play, language, and deferred imitation at this age all suggest the creation of mental images, detached from specific contexts and thus adaptable to many. Such imagery reflects a rapid, continual restructuring of reality as directly experienced, and thus of adaptive possibilities and actions in advance, without the haphazardness of prestructured forms or trials and errors, or mere passive submission. The ultimate power of this inventiveness lies in the fact that future change can now be anticipated, and action devised to preempt, or make change happen. Historically, this seems to have been one of the most prominent aspects of human intelligence.

BIBLIOGRAPHY

Butterworth, G. 1999. *What Counts: How Every Brain is Hardwired for Math*. New York: Free Press.
Ceci, S. J. and J. K. Liker. 1986. "A Day at the Races: A Study of IQ, Expertise, and Cognitive Complexity." *Journal of Experimental Psychology* 115: 255–66. A classic study of complex cognition in context (also referred to in chapter 2).
Chapman, M. 1988. *Constructive Evolution: Origins and Development of Piaget's Thought*. New York: Cambridge University Press. A clear and detailed account of the origins of constructivist thinking as expressed in Piaget's theory.
Dehaene S. 1997. *The Number Sense: How the Mind Creates Mathematics*. New York: Oxford University Press. A model of mathematical intelligence as developments from innate constraints.
Goodwin, B. 1994. *Theoretical Biology: Epigenetic and Evolutionary order from Complex Systems*. Baltimore: Johns Hopkins University Press. A constructivist account of development from a biological point of view.
Greenough, W. T., J. E. Black, and C. S. Wallace. 1990. "Experience and Brain

Development." In R. P. Kesner and D. S. Olton, eds., *Neurobiology of Comparative Cognition*. Mahwah, N.J.: Lawrence Erlbaum. Review of contemporary concepts and approaches regarding relations between experience and development.

Light, P. and G. Butterworth, eds. 1993. *Context and Cognition: Ways of Learning and Knowing*. Mahway, M.J.: Lawrence Erlbaum. Includes representative contributions on a variety of researches suggesting (as Light puts it) that social context is not just a "factor," but constitutional in human intelligence.

Piaget, J. 1988. "Piaget's Theory" (1970). In P H. Mussen, ed., *Manual of Child Psychology*. New York and London: Wiley; reprinted in K. Richardson, and S. Sheldon, eds., *Cognitive Development to Adolescence*. Mahwah, N.J..: Lawrence Erlbaum. Offers a brief overview by Piaget himself.

_____. 1980. *Adaptation and Intelligence*. Chicago: University of Chicago Press. Piaget contrasts constructive intelligence with other mechanisms of adaptation.

Piaget, J., et al. 1972. *Epistemology and Psychology of Functions*. Norwell, Mass.: Kluwer Academic.

Richardson, K. 1998. *Models of Cognitive Development*. Philadelphia: Psychology Press. Offers a critical review of all the different models of cognitive development.

Smith, L., ed. 1996. *Critical Readings on Piaget*. New York: Routledge.

Sternberg, R. J. and R. K. Wagner, eds. 1994. *Mind in Context: Interactionist Perspectives on Human Intelligence*. New York: Cambridge University Press. Another set of "contextualist" views, including the paper by Ceci and Roazzi mentioned in the text.

7

Social Intelligence

In the psychological models reviewed so far, there has been a strong tendency to view intelligence as a property of individuals. This has long been the dominant approach to the description of psychological characters in Western psychology. Psychological characteristics belong to each individual, so it seems natural that the boundaries of the individual should also delimit the extent of the contents and processes to be described, including those of intelligence.

At the same time, there has always been a somewhat subdued recognition that humans are uniquely social, too: that humans act in social conglomerates, and that this might influence the general form or individual differences in their intelligence, or both. Psychologists have responded to this recognition in various ways. One of these has been the traditional resort of arguing that human social intelligence is simply innate. We saw in chapter 3 how recent studies of social behavior in infants has fostered a resurgent nativism. Theorists such as Colwyn Trevarthen of the University of Edinburgh have long been convinced that the presence of infants' social responses and social expressions, even in the first few weeks of life, means that they are "fitted in" from the start. Jerome Bruner has said that social intelligence seems to be "there," as if from conception. In 1992, American psychologists L. Brothers and B. Ring duly proposed that all social information—the processing of faces, voices, and certain movements—is dealt with by a specialized social module in the brain, designed or constrained by our genes.

Another approach to explain and characterize human social intelligence has been to look to other primates for enlightenment. That many other species exist as social groups, presenting to the individual a particularly complex and structured social environment, has long been recognized. Accordingly, authors such as Nicholas Humphrey of Cambridge University have stressed the role of the social world, especially in primates, as the seat of intelligence.

This has been called the "social function of intellect" hypothesis, and suggests that the need to reconcile individual needs with the social cohesion of the group requires a new order of individual smartness or cleverness that is duly reflected in the advanced intelligence of primates in general, and humans in particular. Humphrey, for example, suggests that the range of technological skills, the long period of dependency in offspring, and the complex kinship structures found in primate groups all require extensive powers of social foresight and understanding. American primatologist J. H. Crook and others point to the constant need to read others' intentions, feelings, and states of mind. At least one study has indicated a relation between social complexity (albeit using only the crude index of group size) and individual intelligence in primate groups. Richard Byrne says that this is leading us to a real understanding of primate intelligence.

In order to extend this idea to humans, we would have to assume that the way of life of humans is similar to that of other primates, with similar consequences for their intelligence. Many have accepted that there are close parallels. For example, in his book *The Human Primate*, Oxford neurobiologist Richard Passingham says: "The human strategy for survival is basically the same as that of his close relatives." This, then, all too readily invites a view of primate intelligence in general, and human intelligence in particular, as a process of individual struggle against external social forces, which are in turn viewed as a means to be more or less cleverly manipulated for personal ends. Human social life is thus depicted as a dramaturgical contest, a modern version of the Hobbesian war of all against all, in which those with the sharpest wits, by virtue of their better genes, gain the upper hand.

Another approach has taken into account the fact that humans, unlike other primates, operate in the world through rich cultural institutions: organized forms of production, technologies, shared knowledge, language, beliefs and practices, and so on. But here, again, there has been a tendency to simplify these cultural creations in various ways. One approach has been to suggest that shared cultural practices can each be treated as a group character. The different cultural characters found in different human groups can then be treated exactly as analogues of individual characters, such as height or eye color. Improvement and domination of some over others is then a matter of natural selection, and so can be studied, as various authors have put it, using Darwinian concepts and Darwinian methods. Thus culture is said to change by the chance mutations introduced by individual learning, which then, by selection processes, permeate the culture to replace older forms, according to the adaptationist process described in chapter 3.

In all these efforts to explain and characterize human social intelligence, therefore, we find a seemingly irresistible tendency to reduce its forms and powers to expressions of others already known, namely biological ones. Even human culture is brought within the same, single frame of reference. This seems to be the only way that many theorists can deal with the enormous puzzle that is human intelligence.

An alternative view is that this is not only bad psychology, it is also bad biology. In previous chapters we have seen that, even as accounts of intelligence in subhuman species, they have problems. The theory of hierarchically integrated systems described in this book has suggested how cognitive regulations emerged as a distinct level of intelligence, integrating but opening up epigenetic regulations. These, too, have integrated underlying genomic regulations, which have in turn utterly transformed the intelligence in raw genetic information. In each of these integrations, we see a vastly more powerful system of adaptability emerging by interaction between separately limited levels. All of this gain is lost by collapsing everything back into a single frame of reference.

In the rest of this chapter, then, I want to show how human social intelligence has emerged as a dauntingly powerful force from the interactive relations between a cultural level of social life and the cognitive regulations of individuals. The upshot of such interaction is not just a useful add-on—a useful set of tools to be exploited by the real intelligence within—but a new system of intelligence, incorporating, but transforming and extending, evolutionarily more primitive regulations. Just as all the preceding advances in animal intelligence have consisted of the transformation and extension of more primitive forms, so all the dazzling characteristics of human intelligence stem from this.

Human Intelligence Is Different

Even some primatologists have been of the view that human intelligence is completely different from that of the ape or the monkey. American primatologist David Premack, who has studied chimpanzees in great depth and might be expected to concentrate on the similarities, points out (in an article in 1996) that the human alone teaches. Countless observations with young chimps and other species have shown how they can acquire various adult skills by learning or imitation, but they are never taught. Teaching, on the other hand, seems absolutely mandatory among humans. Such widespread pedagogy appears to be a specific aspect of a more general ability for social cooperation that is also unique to humans. Premack says that a major manifestation of this is an agreement to share, which again seems almost unique to humans: "Even in the chimpanzee where we find hunting and sharing in the sense of physical transfer of goods, we have absolutely no evidence of agreement to share or reciprocation. This is what distinguishes the human development ladder."

Humans are, of course, behaviorally different from all other primates in several other critical ways. They walk bipedally, have complex technologies and other systems of production, and a system of communication radically different from any other in the animal world. But above all else, humans cooperate in their survival and other activities in a way only vaguely prefigured in other primates. Whereas the service of individual needs remains very much a question of individual activity for the chimpanzee, there is evidence even in the earliest human ancestors from more than a million years ago that humans cooperated far more intensively in activities such as group hunting and foraging, the production of tools and shelter, food preparation, defense, and distribution of resources.

The circumstances that fostered the evolution of these abilities are now generally understood. They seem to have presented yet another ratchet upwards of environmental uncertainty and complexity, as the climate dried, forests thinned, and former forest dwellers were forced onto the open savannah or forest margins. Without natural defensive equipment and deprived of traditional food resources, they became extremely vulnerable. The general idea is that these challenges were met by a unique degree and quality of social cooperation. Defense, hunting, and foraging became vastly more effective in an organized group than for a mere collection of individuals. Likewise, while living in larger groups, reproductive relations, child rearing, division of labor and its products, and so on became less fraught when socially ordered. Meeting these new, social demands presented demands on cognition that were far more complex than those experienced in the physical world alone.

Since about the 1990s, another view of the nature of human intelligence has become popular which stresses the nature of human cooperativity as the root of human intelligence. Without a doubt, the most influential theorist in this area has been the Russian psychologist Lev Vygotsky, whose works, written mainly in the 1930s, have only relatively recently been translated in the West (Vygotsky died in 1934). In some of his early work, Vygotsky directly compared the intellectual capacities of children and apes, using the kinds of tasks devised by Wolfgang Kohler and described in chapter 1. These studies showed how intelligence in human children was not only "cleverer," but also qualitatively different, from that of the other apes.

Vygotsky concluded that the difference comes from the way that human thought and activity are embedded in social life from the moment of birth. He argued that the entire course of a child's psychological development from infancy is achieved through social means, through the people surrounding the child. As Vygotsky and his colleague the Russian psychologist A. R. Luria put it:

> In order to understand the highly complex forms of human consciousness one must go beyond the human organism. One must seek the origins of conscious activity . . . not in the recesses of the human brain or in the depths of the spirit, but of the external conditions of

life. Above all, this means that one must seek those origins in external processes of social life, in the social and historical forms of human existence.

In the rest of this chapter, we shall see that engagement in social cooperation accounts for both the quantitative scale of human intelligence and the myriad forms it actually takes.

The Cognition-Culture Complex

As I have mentioned several times, I believe it is the informational depth at which predictability has to be found that best defines the demands on the intelligence of species. Such demands became considerably amplified when our ancestors first started to cooperate. When even as few as two people act jointly over a common task, such as lifting an object, they are obviously changing circumstances, not just passively adapting to them. But, in the process, they also reveal a depth of forces that would not be experienced by either one operating alone. The natural relations between forces of mass, gravity, shape, and friction, and the action of one participant, all become conditioned by the actions of the other. Each has now to take account of that new complexity of forces in order for the joint action to become coordinated. In addition, of course, all these forces have to be geared to an overall, shared conception of joint purpose. In hunting and defensive actions against predators it is even more complex, because the object is itself active and reactive against the joint action. This depth of forces isn't even remotely experienced by noncooperating animals.

I believe that these "representational hyperstructures"—situations in which individuals work together, change the environment, and react accordingly—are, in humans, much deeper and more complex than their counterparts in any other species, because of a unique degree of social cooperation. This is what explains the need for bigger brains. And the refined and extended predictability provided by those structures explains humans' greater intelligence. Even simple cooperation meant that predictability, such as who makes the next move, and what should it be, could be retrieved only from much deeper infor-

mation (covariation) interactions. This, in turn, set in train the virtu-
ous spiral of cerebral expansion, still more intricate social interaction,
sensitivity to hyperstructural information, and so on.

But the consequence of social cooperation was more than just a
capacity gain. The joint regulations necessary to coordinate social
cooperation crystallized in what Vygotsky called cultural tools. These
include the vast range of technological tools fashioned for shared use,
from stone axes to computers and a panoply of psychological tools
through which individuals think and act with others. Human intelligence, then, becomes fashioned by its sociocultural tools. As Vygotsky
explained:

> By being included in the process of behaviour, the psychological tool
> alters the entire flow and structure of mental functions. It does this
> by determining the structure of a new instrumental act just as a tech-
> nical tool alters the process of natural adaptation by determining the
> form of labour operation.

Now let me attempt to illustrate some of the processes through which
social regulations both transform and vastly extend individual cognitions.

Action with Objects

The human world consists of objects, and it has been traditional to
think of intelligence in terms of an individual's knowledge of objects,
and the individual ability to use them. One currently popular view is
that much of our knowledge of objects is innate. Steven Pinker, for
example, says that this ability involves mental programs shaped by natural selection to allow our ancestors to master rocks, tools, plants,
animals, and other people. But in the human social world, even rocks
don't appear in static, persistent form just waiting to be mastered.
Depending on where we are in the world, and in which millennium or
century, a given rock may be an obstacle to be moved, a weapon to be
thrown, a component of a shelter, a stepping stone in a stream, an
object of archaeological or geological study, a source of information

about human evolution or human history, and so on. Each of these requires more than static programs about static objects, and if there is no such constancy there can be no natural selection.

Even a cursory inspection brings home to us the fact that human use of objects hardly ever consists of isolated cognitions around isolated entities. Objects enter into our lives in ever-changing dynamic forms, almost always involving other people. Knowledge of objects, therefore, consists of far more than propositions about its surface appearance and superficial physical properties. In addition, humans are prolific creators of objects as the very tools of their survival, so we have come to live in a world of artifacts.

In these ways, objects are integral to patterns of social relationships, and we know objects—their properties, uses, and so on—through those patterns. This is the special intelligence about objects which children must develop, and which contrasts with the intelligence about objects of other animals. As American psychologist Sylvia Scribner explained, a child's relations with others are largely mediated through objects (a bottle, a spoon, or a chair, say), so the physical properties experienced are those manifested through their social modes of usage.

As mentioned above, social action with objects reveals far more about the deep properties of objects than purely personal encounters possibly could. The more detailed physical properties of objects revealed by cooperative actions around them is what may have first led to their use as tools, defense, shelter, warmth, and so on. The socially embedded knowledge of objects thus transformed and extended individual cognitions, and cultivated an intelligence about objects that is only vaguely foreshadowed in other species.

Concepts

A word such as *chair* or *table* doesn't usually refer to a single item, but to a whole category of items. This general mental representation is what we usually mean by a concept. How concepts are formed has been a major topic in psychological research since from about the late 1970s, motivated particularly by an understanding of the special properties of concepts. By allocation to a category, we can predict far more attributes of an object than we could from current information alone—with obvious advantages for mental economy and intelligence, it is claimed. Little wonder, then, that the American psychologist Jerry Fodor claimed that, "The nature of concepts is the pivotal theoretical issue in cognitive science; it's the one that all the others turn on."

However, although there has been much debate, and many models of concepts have been proposed over the past twenty or thirty years, they have been disappointing because they have been based on the computational principles described in chapter 3. They have almost invariably assumed the presence, in objects being experienced, of clear, identifiable features (such as the wings, feathers, or beak of a bird) as the building blocks of the internal representation or concept. They are just the sort of triggers required by the feature detectors of a computational model (and they can, of course, be easily input at a computer-modeler's keyboard).

What such modeling overlooks is the dynamic social context in which nearly all objects are experienced. All objects are experienced through myriad different orientations, angles, and distances, usually in some sort of motion (even if it is the apparent motion produced by ourselves moving around or manipulating them), and often with parts that are partly or entirely obscured from view. Features do not simply present themselves to the cognitive system in well-structured, repeatable forms, so it is difficult to see how a reliable concept can be formed from them. This is one reason why computational models of concepts have come under considerable criticism recently.

I argued in chapter 6 that we actually form a concept of an object by setting up an internal representation of its deep covariation structure, which is all that remains constant when all else is under constant transformation. It seems reasonable to suggest that the covariation structure of an object, such as a chair, will also co-vary with that of other objects from the same class, so it will easily coalesce into a wider representation of the category as a whole. This, then, is the general representation we tap into when we use a word like *chair*. The same structure will co-vary with aspects of use (as with a chair and the motions of the body used in sitting down), and with those of other objects from different categories, and, most important, the actions of other people, in "event schemas" (such as the regular activity of bringing a chair up to a table, catching a bus, shopping, or going to work). So we end up with what I have called a "hypernetwork," which vastly extends our powers of predictability in the world.

Vygotsky made the point that it is only in the laying open of underlying properties of objects and events, and the need to communicate about them in social interaction, that brings together members of concepts into more abstract relations in this way. In order to transmit our experience of an object to others, he argued, there is no other way than to ascribe the content to a known class, defined by a word, a process which, itself, requires prior generalization. So Vygotsky argued that social interaction and all the powers of generalization go together, the one presupposing the other, and that the greater powers of generalization become possible only with the development of social interaction. In fostering concept formation, then, social regulations radically alter cognitive regulations, and this process unleashes enormous cognitive powers in people as concepts become extended in vast knowledge networks.

Knowledge

Even the cognition of subhuman animals is based on conditioned covariations, as we saw with the barn owl and other examples in chapter 6. But when these are encountered in the course of social cooperation, those conditionings become several levels deeper. This new kind of depth is the chief characteristic of human knowledge generally, so human intelligence reflects, more than anything else, the depth and complexity of covariation structures used in making predictions about the world. All progress in human intelligence, both historically and in individual development, consists of the discovery of a fresh layer of covariation in a domain, interacting with and conditioning those already represented. This structure, in turn, gives structure to cognition. As Robert Glaser of the University of Pittsburgh has pointed out, effective thinking is the result of what he called "conditionalized knowledge"; that is, knowledge that adequately reflects the deep contingencies governing the domain in question.

In everyday knowledge, this covariation structure remains implicit: we are not conscious of it. This has been shown both in studies of "situated cognition" and in recent studies of "implicit learning." For example, in the studies mentioned in chapter 2, variables in a simulated system (such as a factory) were structured according to a complex equation involving interactive terms; in other words, they were based on a system of nested covariations. After training, subjects seemed to have registered these interactions in all their depth and complexity because they were able to make reliable predictions about the system. But they were unable to describe how they were doing it.

However, the frequent need to be explicit about our knowledge, as in sharing, teaching, working together with it, and communicating about it, means turning knowledge into a more declarative or propositional form. Generalizing covariation structures into concepts, each labeled with a word, as just mentioned, is part of this process. Through speech and writing, these more explicit knowledge structures can then be reworked and extended into more complex forms, which are further interrelated, used as metaphors for exploring new domains, and so on. This process has reached its most sophisticated form in the cultural tool we call science.

Science

Implicit knowledge, despite being structured within a social framework, is made explicit only when we need to communicate about it in social discourse and social activity. It is these deeper layers of covariation structure that are exposed by progress in scientific knowledge. Indeed, it is the task of scientific research to discover and describe, using the most precise terms possible, the underlying structure of the natural world, as distinct from its surface appearance. Scientific discoveries expose the previously unsuspected, deeper interactions among variables governing a domain or phenomenon. Sometimes, of course, this process can radically alter our interpretation of everyday observations, such as whether the Sun goes around Earth or vice versa. In this way, predictability of future states from current conditions and possible interventions becomes ever more precise.

Often our failure to have fully achieved this predictability reflects a limited grasp of the depth of the interactions involved. We know, for example, of the many variables associated with heart disease: smoking, diet, exercise, stress, and so on. But puzzles remain because many people, despite scoring high on most if not all of these variables, do not exhibit heart disease. What we need to know are the deeper interactions between these, and other factors, in the sense just mentioned.

The method by which scientists have come to discover scientific knowledge is one specially designed for the painstaking explication of just such deeper covariations in nature. This is one of the most impressive, and most intelligent, examples of the way that social regulations have appropriated and vastly extended otherwise limited cognitive regulations. It is now well documented how the great flowering of science from the seventeenth century onwards entailed a new methodology: making observations more systematically, constructing explicit theoretical models, making predictions (hypotheses) from them and testing them in controlled experiments. This is what MIT psychologist Susan Carey has called a complex "logic of confirmation."

What happened in the seventeenth century was a process in which natural philosophers, previously working independently (and now very much with the encouragement of their monarchs and new economic conditions), started to probe nature with manipulations, to test their theories with controlled experiments, and to present their work for the reflection of others. The need for every step to be made explicit, so they could be replicated by others, brought out the logical requirements of modern empirical science. That is, the current methods of science grew naturally out of joint reflection. Indeed, the American linguist Philip Lieberman has argued that the upsurge in the social processes of conferencing, cooperation, and publication among scientists in the seventeenth and eighteenth centuries was the single most important step in the historical development of the scientific method. The logic of confirmation, in other words, also appears to be an example of the transformation and extension of otherwise limited cognitive abilities by regulations at a higher social level.

Reading, Writing, and Memory

As we have seen, joint action on the world lays open far more of its underlying complexity than does the solitary action of an individual.

In addition, joint attention and communication about objects and events brings about generalization in the form of concepts. These concepts then became socially marked in the agreed, but transitory, sounds of language (of which more shortly). At some time in human history, around ten thousand years ago, they also began to be marked in another way, by more permanent, visible marks: writing. The origins of this new cultural tool have been traced to the settlement and population expansions following the agrarian revolution about ten thousand years ago. This revolution led to increased specialization, production of surpluses and trade, and the need for better systems of record-keeping.

Even the first of these new tools of representation, such as marks on sticks and, later, pictographs, furnished not only a crucial medium of social cooperation but also of extended cognition. Perhaps this is best seen in their effects on memory. The strongest tradition of memory research has consisted of having people remember and immediately recall nonsense syllables, with all meaning and purpose stripped away. The main objective of this research seems to have been to establish the inherent cognitive principles of memory. But it results only in a description of a degenerate form of memory. As Vygotsky and Luria pointed out, the written forms of communication vastly expanded the memory function, transforming the natural function in the process and furnishing a new medium of cognitive organization and planning. It is this superimposition of the cultural form over the natural form of memory that constitutes the most crucial ingredient of memory development in children, they said. Such auxiliary memory tools have, of course, been vastly augmented in more recent times by printing, libraries, calculators, and computers.

Number

Number (along with logic, of which more later) is often considered to be the highest creation of human intelligence (and those who use it tend to be thought of as the most intelligent individuals). Yet number, too, can be described as a socially devised tool for explicating, for the purposes of communication and cooperation, the deeper relations in nature. In the process, it vastly extends our otherwise limited cognitive abilities and representational formats.

In his book *The Number Sense*, Stanislas Dehaene suggests that animals, as well as newborn humans, have a rudimentary number competence. He says this is a primitive number module, evolved by natural selection, coded in our genes, and possibly located somewhere in our brains. However, he also says (somewhat contradictorily) that this does not amount to "a digital or discrete representation of numbers." As we saw in chapter 6, it seems to make more sense to see this primitive ability as simply the discrimination of variation at a sensory or perceptual level. The specific ability of a conceptual system is to capture the covariations (often lying very deep, in the sense also described in chapter 6) that render a changeable world more predictable.

A concept of number arises from covariations among quantitative variables across different categories of objects, independent from the identities of the objects themselves. For example, the change from one to two to three eggs co-varies with the change from one to two to three spoons (or whatever). This is different from mere quantitative discrimination. A blackbird may be able to discriminate three worms from four, but only if we saw it collecting precisely three worms for three chicks simultaneously could we attribute it with a number concept. As the British philosopher A. N. Whitehead noted: "During a long period, groups of fishes will have been compared to each other in respect of their multiplicity, and groups of days to each other. But the first [person] who noticed the analogy between a group of seven fishes and a group of seven days made a notable advance in the history of thought." However, I think that the crucial spur to such comparing and noticing in humans—and with it the all-important digitization and symbolic marking of quantities—was the unique degree of social cooperation in human activity.

We know from the history of written number notation that such digitization of quantities into a communal system of symbols, signifying increments and totals, has been around since at least twelve thousand years ago. At that time, they consisted mostly of marks on sticks, knots on twine, pebbles, and so on. But sequences of notches on bone, indicative of something being counted, have been discovered from human habitation sites as old as twenty thousand years or more. We must presume that the oral symbols "one," "two," "three," and so on must have been in use well before that. It is difficult to think of a purpose for the invention of such symbolic utterances or markings other than that of human communication. Such communication about quantities would have already been important in hunting and gathering communities (signaling about the number of prey in a group, the number of predators in a woods, the number of tools needed for digging, and so on). But that importance must have been vastly inflated with the agrarian revolution after the tenth century B.C., which involved long-term settlement, productive specialization of social groups, and vastly extended trade between them.

So animals and infants have a quite sophisticated capacity for quantitative discrimination, and even for detection of complex covariation. But it was the need for human communication, itself driven by social cooperation, that directed the attention, resulting in the digitization of the covariation we call number. Such digitization presumably evolved by social negotiation and ultimate agreement over a long period of time. It was obviously easier with quantities such as sets of objects, whose increments are already discrete, than with continuous quantities, such as grain and water, where units had to be socially defined and agreed. But no human group has been without such a system, although their diversity of form is quite astonishing. For example, some groups have used body parts as the number symbols, and various bases (other than our own base ten) have been adopted for expressing large numbers by embedding them hierarchically.

But number as we know it today is more than a system of marks or symbols. The Ancient Greeks were struck by the seeming mystical qualities of number, sensing some deeper properties within. We still find this awe and mysticism about number. It is explained by the fact that number is more than a conventional set of tokens. Any use of number unavoidably implies the deeper, complex covariations on which a number system is based. Piaget has pointed out how the concept of number itself coordinates two deeper concepts. First, the number concept involves sets in class-inclusion coordinations: two is included in three, which is included in four, and so on. Second, the sets can be put into series in space or time: three is greater than two, which is greater than one, so three is greater than one. Without these concepts, the logic that is so fundamental to a number system could not be used, and even simple arithmetic, such as addition or multiplication, would not be possible. Every child needs to understand these logical necessities before the number system can be properly understood.

Once a number system had been established in its digitized form, the underlying logic allowed it to be further manipulated to create all the arithmetical and mathematical constructs that amaze us. The Ancient Greeks were dazzled by the hidden logical structure in number which permitted this, believing that therein lay the Divine essence of the whole universe. They thought they had discovered a mystical property, or "higher unity," that generated all the other combinations of number making up what was called the *kosmos*, meaning "good array."

Perhaps the real significance of the manipulation of number, or mathematics, is the way that it helps us achieve an explicitness about the deeper relationships in nature that are not attainable by ordinary language. As Whitehead put it: "The originality of mathematics consists in the fact that in mathematical science connections between things are exhibited which . . . are extremely unobvious." Accordingly, in the seventeenth century, mathematics became the language of science, eventually taking us from a static to a dynamic world with the invention of calculus.

Mathematics is therefore another way in which formerly limited (quantitative) abilities have been brought out and elaborated by the demands of joint attention and joint action. As Piaget insisted, becoming intelligent in mathematics involves understanding certain logical relations. But these are embedded in social use, which is why many contemporary mathematics educators are stressing the need for maths to be learned in meaningful social contexts. When acquired in the context of its social use—in contrast with the social detachment of most learning of mathematics in schools—mathematics can be transformed from a dreary subject to a lively tool readily mastered by all, according to Terezinha Nunes of London University and the American psychologist Jean Lave, among others.

It seems strange today that number intelligence is so often accorded the same innate, metaphysical roots as it was by the Ancient Greeks. It seems equally strange that so much psychological theory about cognition and intelligence is still in the era of computation on static entities, having not yet moved from an Aristotelian to a dynamic conception of the world, as happened to number in the seventeenth century. But, as we have seen, number is not an innate module or mental organ, but a cultural tool, devised out of social cooperation and social discourse, and it vastly extends our otherwise limited cognitive regulations.

Logic

If mathematics is the use of symbols of number to realize and demonstrate necessary numerical truths for social purposes, the study of logic has been the search for a system that can do the same for truths in general. Logic (however we describe it) has thus long been considered to be the very epitome of mature intelligence, as intrinsic to the cognitive system as the procedures of filtration and excretion are to the kidney. Yet, as we saw in chapter 6, such inherent logic has been extraordinarily difficult to find and describe.

Many studies have shown how logic is another example of the way that social regulations between people come to define and extend the cognitive regulations within them. This has become most clear in studies using verbal syllogisms, sometimes used, indeed, as tests of a person's logical abilities. For example:

All the boxes from room A contain cups
This box contains cups
Did it come from Room A?

That the ability to perform such logic consists of rather more than a simple mechanical process became clear when the effects of context and/or background knowledge were pointed out. As we have seen, most people err when given the following problem:

All A's are B's
Are all B's A's?

But not when the same logic is presented as follows:

All humans are mammals
Are all mammals humans?

A natural inclination may be to think that "clever," logical people can perform the more abstract version of this task because of some greater brain power. But cross-cultural studies have shown that judging what is or is not logical thinking is not a simple, objective process because the logic comes in socially determined forms. For example, in research in the 1970s Sylvia Scribner tried to administer logical syllogisms, such as the following, duly couched in local cultural terms, to Kpelle farmers living in Nigeria:

All Kpelle men are rice farmers
Mr. Smith is not a rice farmer
Is he a Kpelle man?

But the first subject just kept replying that he didn't know Mr. Smith, so how could he answer such a question?
The following is another example:

All people who own houses pay a house tax
Boima does not pay a house tax
Does Boima own a house?

And the answer was: "Boima does not have money to pay a house tax."
Are such subjects simply lacking in logical abilities? After all, Scribner showed that Western schoolchildren can answer these questions quite easily in the expected logical manner. This turns out to be a premature conclusion. As British psychologist Philip Johnson-Laird has pointed out, the subjects in both the examples just cited are actually making perfectly logical deductions from premises. For example,

All the deductions that I can make are about individuals I know
I do not know Mr. Smith
Therefore I cannot make a deduction about Mr. Smith

The difference lies in the use of different, socially devised, logical forms. The difference, says Scribner, is that the standard syllogism depends on using only information that was provided by the tester, and so is internal to the problem. This is typical of school-type tasks in which pupils are disciplined to reach foregone conclusions from fixed premises. It reflects not so much a superior logic as a specific set of social relations, entailing specific assumptions about what information is to be activated or nonactivated in deducing the conclusion. The standard logical test is a kind of game with tacit rules requiring the subject to ignore information outside the premises given. Unlike Western subjects, nonschooled subjects tend to bring in all their background knowledge, as well as that of current context, in reaching an answer that is at least as logical.

Whether or not someone is assessed as having logical ability, therefore, turns out to be a question of cultural assimilation of psychological tools. Patricia Cheng of Carnegie-Mellon University and Keith Holyoak of the University of Michigan have argued that all logical reasoning takes place within abstract knowledge structures induced from everyday social experience. They call these structures "pragmatic reasoning schemes." Perhaps not surprisingly, in view of the above discussion, they have shown that extensive training in abstract logic does not improve performance on logical tasks, whereas even brief schema training, pointing out how the task could be related to an everyday situation, improves performance significantly.

This view of logic can also shed further light on the cultural specificity of IQ tests. For example, we saw in chapter 6 how versions of Raven's matrices, based on exactly the same logical structure but couched in terms of feasible social situations, transformed children's performances. Why should this be so, when the traditional Raven's items are considered to be culture-free and thus a pure measure of inductive and deductive logic?

The answer, of course, is that they are not culture-free. It isn't difficult to show that Raven's matrices are culturally structured tasks. First of all, the items are administered in a socially fraught test situation, itself a specific cultural device. Then they are presented as black-and-white figures, flat on paper, with the order of information arranged from top-left to bottom-right. These are culture-specific tools for handling information, and are more prominent in some groups than others.

Similar criticism can be applied to the logic embedded in the items themselves. These require the induction of the logical rule implicit in the arrangement and transformation of the visual patterns on the paper, in order to be able to predict the form of the one missing from the bottom row. For example, one rule simply consists of the addition or subtraction of a figure as we move along a row or down a column (see figure 2.2 in chapter 2 for an example); another might consist of substituting elements. My point is that these are emphatically culture-loaded, in the sense that they reflect further information-handling tools for storing and extracting information from text, from tables of figures, from accounts or timetables, and so on, all of which are more prominent in some cultures and subcultures than others.

In other words, items like the Raven's matrices are rather more than tests of abstract logical powers: they contain hidden covariation structures, which make them more, not less, steeped in culture than any other kind of intelligence-testing item. As American psychologists D. P. Keating and D. J. Maclean put it, they are "the most systematically acculturated tests." This culture specificity is hardly surprising because, as already mentioned, the items are the products of the cognitions of human beings, themselves immersed in a specific cultural milieu. The idea that the Raven's matrices somehow separate children on the basis of a general logical power, intrinsic to the child, rather than simply screening them for cultural background, seems quite false.

Language

Cooperation, and the meshing of multiple actions, requires communication. Language translates the deep structures of cognitive representation into linear, temporal patterns of sound and visual symbols interpretable by others. Unfortunately, for most of the twentieth century, and especially since the 1950s, language has been analyzed, like cognition in general, as a computational system, with its innate core principle or set of rules. Sentences have therefore been pulled apart as parcels of sound and meaning in an attempt to discover the self-sufficient rules by which they were put together in the first place. No full account of them has yet been offered or agreed upon.

This was the movement initiated in the 1950s by Noam Chomsky. In more recent times, Pinker has spoken of the "language instinct" and insists that "the capacity for language is part of human biology, not human culture." This is not the place to enter into this debate into detail. Suffice to say that the stock arguments about innate language, such as the presence of language universals (or similar language structures in widely different cultures), the argument from the poverty of the stimulus, and so on, appear compelling but have been repeatedly challenged. What is worth challenging here is the whole mode of analysis that views language as a computational system within the brain; this is what leads to the debate in the first place. I have little doubt that if we analyzed the multitude of hand movements we do in a day in purely structural terms, abstracted from their social and cultural purposes, we could arrive at the reasonable-sounding hypothesis that they were governed by an innate hand-movement program!

Fortunately, there is another approach to the study of language, which views it (rather as we have just done with logic) in terms of interactions between an internal capacity and the social regulations that extend and shape it. Foremost in putting forward this view has been the Russian linguist Mikhail Bakhtin. Because speech is used for communication in human actions, Bakhtin eschewed the analysis of speech as self-contained linguistic structures abstracted away from their purpose. He saw it as a form of action in social context. Rather as I spoke of different logic games above, so Bakhtin drew attention to the way that speech exists in genres, as culturally determined forms with typical social purposes (greetings, commands, requests, guidance, and so on). Just as our multifarious hand movements (at the computer, the table, the piano, or the lathe) assume forms that can make sense only in their cultural context, and never as self-contained computational rules or principles, so we can understand language only as forms of social action or genre. As American psychologist James Wertsch and his colleagues have suggested, an utterance is invariably carried by a specific genre, just as it is expressed by a national language, such as English, French, or Thai.

In *How Brains Think*, theoretical neurophysiologist William Calvin sees language as a foundation of intelligence. But it is better, in my view, to see it as a tool or instrument of intelligence: as an expression (and amplifier) of it, but not its cause. In this view, then, language is another cultural tool through which social regulations mediate the cognitions of participants in the culture. The way in which this happens has frequently been discussed. Piaget noted how the development of symbolic functions in the second year of life (play, mental images, drawing, and, above all, language itself) "enables the sensorimotor [intelligence] to extend itself." Vygotsky claimed that, once children learn to speak, their behavior becomes entirely different from that of other animals. Speech, by manipulation of its structure (as in thinking verbally), can be used to play a specific part in organizing thought, often giving rise to novel ideas. Thus (as Vygotsky put it), we often come to solve a practical task "with the help of not only eyes and hands, but also speech."

The Dialectics of Intelligence

It has often been said that human intelligence is "greater" than that of all other animals, but this has been explained, rather unsatisfactorily, as a kind of difference in capacity. Often the impression created has been that our brains accidentally got bigger, affording us the intelligence with which to make our more complex social arrangements. At the same time, individual or group failures in social success and status are explained by lack of intelligence, itself due to lack of genes for the brain structures required, although the identity of these has remained mysterious.

In a sense, the purpose of this chapter has been to show how both of these mysteries are solved at once by the realization that human social life is a whole new system of regulations that take over the cognitive and other regulations already evolved, vastly transforming, shaping, and extending them. Human intelligence therefore resides in what I have called a cognition-culture complex. As Mervin Donald of the University of Ontario put it in his book *Evolution of the Modern Mind*:

> Our genes may be largely identical to those of a chimp or gorilla, but our cognitive architecture is not. . . . Humans are utterly different. Our minds function on several new representational planes, none of which are available to animals. We act in cognitive collectivities, in symbiosis with external memory systems. As we develop we reconfigure our cognitive architectures in nontrivial ways.

The first two or three million years of this interaction consisted of the coevolution of its partners. Emerging social regulations became the context for further growth of the cerebral cortex, which in turn provided the apparatus for grasping ever deeper social contingencies, and so on. The consequences of this new system of intelligence have, of course, been staggering. Since it peaked about a hundred thousand years ago, humans have spread quickly around the globe, occupied every possible niche, devised rapidly changing and developing technologies, and now exist economically as a complex international network encompassing virtually all individuals.

It is important to stress how these powers arise as emergent properties from the interactions between levels, creating a final level of regulation incorporating all those evolved before (figure 7.1). Social regulations arise inter-psychologically, and become internalized, as intra-psychological forms, at the cognitive level, but not in any passive sense. The cognitive level (and its representations) is its own system, and reacts with those at the social level, feeding back new abstractions, and often helping to develop and transform it, in an ongoing dialectical relationship.

In the relations between individual and social levels, this process explains both the contributions of individual minds and the sometimes rapid, or revolutionary, process of cultural change and progress we have seen throughout human history. Its lifelong developmental nature is seen particularly in children, whose minds, as Vygotsky was keen to stress, must not be thought of as being passively programmed by social forces. "The very essence of cultural development is in the

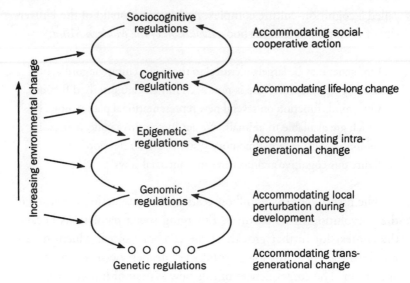

FIGURE 7.1. Cognitive (and other) regulations are functionally extended by being nested in social regulations.

collision of mature cultural forms of behaviour with the primitive forms that characterize the child's behaviour," he said. In this way, the cultural order is, in a sense, reconstructed in each developing child, but often with novel individual forms that can reflect back on broader cultural change. This also explains the vast diversity of human intelligence, between individuals, between groups, and between historical periods.

The upshot is that, far from human intelligence being merely an adaptation, as subject to Darwinian laws as an eye or a leg, humans adapt the world to themselves rather than vice versa. The need to grasp the deeper hyperstructural relations entailed in social action, which expanded the brain, also brought new powers of foresight and insight, and even methods for obtaining them. Humans don't have the epigenetic regulations that result in wings and flying, but we fly better than any bird, thanks to our deeper understanding of physical forces (thanks, in turn, to the social processes of science). Only other social forces—ideological and political—prevent its application for the solution of myriad other problems currently facing humanity.

BIBLIOGRAPHY

Barsalou, L. W. 1993. "Challenging Assumptions About Concepts." *Cognitive Development* 8: 169–80. Typical Barsalou critique of problems in contemporary (computationalist) models of concepts.

Changeux, J.-P. and J. Chavaillon, eds. 1996. *Origins of the Human Brain.* New York: Oxford University Press. Some very readable "state of play" reviews on human brain evolution. Also includes the discussion on the uniqueness of human intelligence by David Premack.

Clark, K. and M. Holquist. 1984. *Mikhail Bakhtin.* Cambridge: Harvard University Press. A readable account of Bakhtin's theory of language.

Humphrey, N. 1988 (1976). "The Social Function of the Intellect." In R. Byrne and A. Whiten, eds., *Machiavellian Intelligence: Social Expertise and the Evolution of Intellect in Monkeys, Apes, and Humans.* New York, Cambridge University Press. Humphrey's original inspirational paper reprinted with more recent views.

Scribner, S. 1997. "Mind in Action: A Functional Approach to Thinking" (1977). In E. Tobach, L. M. W. Martin, R. J. Falmagne, A. S. Scribner, and M. B. Parlee, eds., *Mind and Social Practice: Selected Writings of Sylvia Scribner.* New York: Cambridge University Press. A collection of Scribner's erudite thinking and original investigations, bringing human intelligence to life, as it were, in everyday work and other cultural activities.

Stigler, J. W., R. H. Shweder, and G. Herdt. 1990. *Essays on Comparative Human Development.* New York: Cambridge University Press. Views of culture as constitutional to intelligence, including Jean Lave's paper on math (and other) learning in social context.

Trevarthen, C. 1974. "Conversations with a Two-Month-Old." *New Scientist* (May 2). A very readable paper reflecting the "social-nativist" view that infants have an innate social intelligence.

Wertsch, J. V., F. Hagstrom, and E. Kikas. 1996. "Voices of Thinking and Speaking." In L. M. W. Martin, K. Nelson, and E. Tobach, eds., *Sociocultural Psychology: Theory and Practice of Doing and Knowing.* New York: Cambridge University Press. A paper that contrasts the Western tradition of "individualized" intelligence with intelligence in sociocultural context (the latter is also represented in a number of other interesting papers in this collection).

Vygotsky, L. S. 1962 (rev. ed., 1986). *Thought and Language.* Cambridge, Mass.: MIT Press. Some of Vygotsky's original research and interpretations on the origins and nature of concepts.

_____. 1981/1988. "The Genesis of Higher Mental Functions." In J. V. Wertsch, ed., *The Concept of Activity in Soviet Psychology*. Armonk, N.Y.: Sharp. Reprinted in K. Richardson and S. Sheldon, eds., *Cognitive Development to Adolescence*. Mahway, N.J.: Lawrence Erlbaum. The classic Vygotsky paper on the origins of human intelligence in social relations and cultural tools.

Whiten, A. and R. Byrne, eds. *Machiavellian Intelligence*, vol. 2: *Extensions and Evaluations*. New York: Cambridge University Press.

8

The Intelligent Brain

Nearly everyone believes that intelligence resides in the brain. In the minds of the general public, as in those of many psychologists, intelligence is virtually synonymous with brain power. It has long been thought that thorough examinations of the brains of distinguished individuals would reveal precisely where that power lies (and possibly what it consists of). So we came to be told of the sizes of the brains of Oliver Cromwell, Lord Byron, Franz Joseph Gall, and Anatole France, and how Gauss's brain appeared to have more convolutions than that of an ordinary worker, and so on. In his book *The Mismeasure of Man*, Harvard paleontologist Stephen Jay Gould tells us how, at the height of the fashion for phrenology in the nineteenth century, many renowned scientists bequeathed their heads to science to help identify the roots of greatness.

Much of the comparative approach described in chapter 1 was motivated under the implicit assumption that, by contrasting human brains with those of other species, of lesser intelligence, we can discover what it is that gives us so much more of it. In addition, new means are constantly being devised in attempts to locate specific intellectual functions in specific cortical areas or architectures. Just as psychologists seem to heave a sigh of relief when they feel they can attribute something to a genetic code ("Whew, I'm glad that's that off my mind"), so they sometimes seem to be in similar haste to find this or that brain center which does just what they are trying to describe, as if the burden of further description was then lifted. Finally, much of the delving into artificial neural networks, as in the recent connectionist zeal, is motivated by the assumption that, within the properties of networks themselves, we can discover the mechanisms of human intelligence.

Despite these many speculations, progress in understanding the relations between the brain and intelligence has not been substantial. The past two or three decades have produced an outpouring of interesting new facts and ideas. But it has to be admitted that psychologists and neuroscientists have not managed to weave these into a clear, or clearly agreed upon, theory about how the brain produces, or is otherwise involved in, human intelligence. One manifestation of this backwardness is the overt puzzlement and exasperation sometimes expressed about what our huge brains, and our superior intelligence, are actually for. IQ theorists remain convinced that they are measuring some fundamental cerebral power, but they have still failed to provide a generally accepted description of it. Others have attempted to confine intelligence to specific cerebral modules, with similar lack of detail about what these are, as we saw in chapter 3. In her book *The Human Brain: A Guided Tour*, Oxford neuroscientist Susan Greenfield repeatedly reminds us of the "tantalizing mystery" surrounding relatively simple processes such as recognizing familiar objects. Similarly, in *Rethinking Innateness*, cognitive scientist Jeffrey Elman of the University of California and his colleagues remind us of our near-ignorance about brain structures that underlie language, art, mathematics, music, and "all the complex skills that make us special."

In my view, a major stumbling block to the pursuit of these problems is a poor grasp of the basic functions of higher processes, such as what information the brain uses, and what for. Although it is usually assumed that intelligence is information processing, and the brain is the information processor par excellence, the nature of the information processed is never described, except in very general terms. This problem haunts both neurobiological and psychological inquiry. In most of this chapter, therefore, we shall see what the discussion in the previous chapters (especially chapters 6 and 7) suggests about the nature of the information that the human brain deals with, and thus in what sense it is involved in human intelligence. Before doing that, though, let us take a quick look at the range of alternative hypotheses regarding the relation between brain and intelligence.

Brain Size

A brain-centered view of human intelligence has led to the obvious, if crude, hypothesis of a relation between brain size and intelligence. It turns out, though, that even a simple correlation between brain size and any measure of intelligence appears to be very difficult to establish. A convincing causal association seems to have become increasingly remote. The main reason for this is that the area is riddled with methodological difficulties and inconsistencies, some of which were discussed in chapter 3.

In addition, such a simple hypothesis seems naive in the light of other observations. For example, it is well known that brain volume in females is some 15 percent less than that in males on average, with no agreed upon difference in intelligence. Also, it has long been known that humans with very small brain volumes (less than 900 cm^3), or relatively small amounts of cortical tissue, can be functionally perfectly normal. So whatever it is that creates differences in human intelligence (and thus whatever human intelligence actually is), it becomes difficult to reduce it to straightforward brain size.

Much the same applies to cross-species comparisons. The brief course of human evolution has witnessed a tripling in brain size over that of our nearest primate relatives, but we have only very general notions as to the advantages that are conferred by this "extra" brain size. This still presents a major challenge to theorists. Richard Byrne, for example, offers interesting suggestions regarding the evolution of brain size and intelligence in mammals in general and primates in particular, but then balks at the need to continue such an account for the great leap represented by human brain size and intelligence.

Localization of Function

Instead of gross size, some investigators have pointed to particular brain areas or structures as being the seat of intelligence, or at least particular aspects of it. Suggestions have come from many kinds of observations. The most long-standing source of information has been that of accidental brain damage in humans. On that basis, for example, it was known in the nineteenth century that damage to certain parts of the left hemisphere of the brain was associated with impaired speech and language, so they duly became known as "speech areas." Damage to the corresponding parts of the right hemisphere appeared to be related to spatial deficiencies, such as an inability to arrange small blocks in a certain order (an item in many IQ tests). It has been suggested that the inferior parietal cortex, a small region in the rear half of the brain, is responsible for mathematical thought or other quantitative transformations. Damage to the frontal lobes has led to suggestions that they are responsible for planning and organizing sequences of action, or executive functions, as they are known.

Two other sources of evidence from animals (usually monkeys) have suggested a number of possibilities. These are surgically applied lesions to well-defined areas of the brain, and recordings from very fine electrodes inserted into single brain cells, so that their responses to particular kinds of stimuli can be observed. In these ways, the left temporal lobe, just above the ear, has been implicated as the site of recognition and categorization of objects. Recording from single cells in that area finds neurons that are sensitive to very specific objects, such as hands or faces. Lesions in the same area seem to make it difficult for the animal to distinguish between two objects of different shape. In contrast, lesions to the posterior parietal area seem to make it difficult to learn discriminations based on the spatial arrangement of objects (for example, choosing which of two objects is closest to a third).

Differences in the degree of formation of brain structures is often said to explain individual or group differences in intelligence. American neuroscientists J. Levy and W. Heller conclude in a review in the *Handbook of Behavioral Neurobiology* (1992) that "gender differences in hemisphere maturation, established during the first year of life and present throughout development, condition how boys and girls process incoming information." Sex differences in performance on certain spatial reasoning tasks, and with it school performance in mathematics, have been attributed to differences in brain structure, attributed in turn to hormonal and sex-linked genetic differences. It is worth mention-ing, however, that in Britain at least, the historic differences in mathe-matical performance between male and female pupils, on which such speculation is based, has been completely reversed over the past decade or so. This suggests a sociopsychological, rather than a biological, cause.

Similarly, correlations between the time at which cerebral struc-tures develop and the emergence of specific cognitive abilities have been taken to implicate each structure as the seat of a particular abil-ity. For example, it has been proposed that development of the frontal cerebral lobe in early infancy directly determines "object perma-nence," which is the ability to remain aware of the existence of an object in spite of changes in its appearance, or even its complete dis-appearance. As described in chapter 3, it has been proposed that spe-cific parts or modules of the brain act as intelligence organs.

Perhaps the strongest spokesperson for a brain-centered view of intelligence is American neuroscientist Michael Gazzaniga. In his book *Nature's Mind* (an obviously provocative title), he speaks of "spe-cific circuits for intelligence." Largely based on tests with human patients who have undergone surgical lesions, as well as correlations between size and performance in certain lobes, he concludes that "the evidence is fairly strong that specialized circuits in the left brain are managing the complex tasks of human intelligence."

Recent techniques for scanning the relative activities of different parts of the brain in conscious humans have created much excitement about what goes on where. For example, positron emission spec-troscopy (PET scans) or magnetic resonance imaging (MRI scans) have been used to show changes in blood flow during performance of intellectual activities, such as reading or doing simple calculations. EEG recordings have also been used to plot the time course of the brain's electrical activity during such tasks.

Applying these techniques produces results which are fascinating, though not without problems. The scans represent summations of activity in an area over periods of up to a minute, and so have poor temporal resolution, whereas EEGs are not anatomically well-defined. Although investigators will often stress the diffuse nature of complex function across numerous cerebral sites, the temptation to attribute a given area with this or that aspect of intelligence is sometimes difficult to resist. Thus Stanislas Dehaene noted how some of the recent scanning work is conducted in a "neo-phrenological" framework.

All of this work is, of course, laudable and exciting, but it is subject to misinterpretation or even distortion. The impression often created is that intelligence is a product of the machinations of the brain, or even particular areas of it, and it is now up to neuroscientists to tell us what that intelligence is and how it is brought about. When there is a difficulty—as with a child backward in school, for example—the brain is increasingly becoming the first, rather than the last, candidate for diagnosis. If we come across someone with superior abilities in a domain or subject area, why not do a scan to tell us which brain mechanisms are responsible? The same applies to the explanation of group differences (as in gender differences, for example).

At a theoretical level, developmental and other psychologists are increasingly looking to neuroscientists to offer the models of intellectual functions which their own efforts have so far failed to provide. This strikes me as ironic because, as just mentioned, the confusion among neuroscientists suggests that the very opposite may be the case: it is better psychological theory about intelligence that is needed to provide a firmer framework within which neuroscientists can work.

Let me try to rationalize this situation. Although it is difficult to challenge the general view that different areas of the cerebral cortex are involved in intelligence-related functions, we need to be clearer about what "involved" means. The direction of cause from correlation is not a foregone conclusion. Many commentators have warned about the overinterpretation of functional localization from lesion data. First, it is well known that activities supporting cognitive functions are highly distributed in the brain, and this can create much confusion. Like the mental properties of different social groups, it is possible that these characteristics may not be intrinsic to the separate groups at all, but originate in the relations between them.

Second, even when there is a correlation between a specific area of the brain and some cognitive function, we need to exercise care over the functional attributions we make. Terms used to describe this or that area as specialized for language or specialized for math are only shorthand expressions for the correlation itself, and do not tell us what actually goes on there. As described in chapter 3, it is now thought that what were called speech areas are actually specialized for a much more basic function that happens to be intrinsic to speech processing: the discrimination of changes in stimuli occupying very short periods of time. By the same token, when a recording from a neuron indicates responses to a particular object in its sensory field, this does not necessarily mean specialization for the object in the sense of it being a feature detector: the neuron could be reacting to some much simpler physical properties of the object, such as certain spatiotemporal aspects of its form that contribute to recognition of the object.

Even the classic case of hemispheric specialization (left for language, right for spatial) turns out to be probabilistic and subtle, rather than absolute; in some cases, comprehension and other symbolic abilities are more affected by right- than left-hemispheric lesions. In his book *The Decline and Fall of Hemispheric Specialization*, Australian neuroscientist Robert Efron complains about the "profound epistemological and conceptual confusions that are proving fatal to the field," and suggests, on the basis of wide review and new empirical studies, that the differences may be a much more basic one, involving slight variation in the time course with which sensory inputs are being scanned.

An alternative view is that, far from being determined by a localized architecture, more distributed functions (themselves emerging in interaction with complex and changeable external demands) might use particular cell groups because they have processing properties or connectivities conducive to them. These areas are then further developed and transformed by the function. However, this arrangement is not cast in stone and can be reversed, as long as development has not proceeded beyond a critical point. There is much evidence for that view.

For example, sensory inputs diverted from their usual target area to another by surgical rewiring—say from the retina of the eye to what is usually the auditory area of the cerebral cortex—turns the properties of the latter into those of a perfectly functioning visual area. Cortical tissue transplanted from one area to another takes on the properties of its new location. Removal of sites usually associated with specific functions can have their functions completely taken over by another site, often some distance away from it. And, in children at least, functions associated with the site of a lesion can be replaced or rebuilt by co-opting other networks; they are "anatomically reprogrammed," as French neuroscientists A. R. Lecours and Y. Joanette once put it. All these cases provide strong evidence of the potential for functional reorganization of the developing human brain, and prompt the question of why an organ so evolved has the potential for such functional reorganization intrinsic to it.

Against this background we have to remember that neurological specialization of the cerebral cortex is both limited and confusing. The cerebral cortex shows a remarkable uniformity of structure, constitutionally, across individuals and across all mammals. It has the same cell types, the same six-layer structure, the same broad patterns of connectivity, and so on. Apart from size, there is little that is unique in the details of neurological structure that accounts for human intelligence. Rather, it seems to have been the vast expansion of a basic processing capacity for use by external organizational regulations that appears to define the role of the brain in human intelligence.

Hierarchies and Control

Finally, we come to the idea that intelligence resides in some overall executive function of some part of the brain, and oversees a hierarchy of lesser cerebral functions. Agents at the lower level do little jobs, or deal with particular aspects (such as sort by shape), the results of which get passed on to the next level, which combine these into slightly bigger aspects, and so on, with each level of contribution being bound into the next, until an ultimate product (such as an image) is obtained. This is then dealt with by an ultimate level of decision and responsibility.

Such a hierarchical view is still the most common model of intelligence among IQ theorists, in which special abilities progressively combine into the overall general ability, or *g*. A hypothetical executive function still forms the mainstay of models of cognition and memory, as discussed in chapter 4. Convergent association of information from simple feature detectors into those with more complex and specific sensitivities is still part of neuroscience doctrine. As with genetic determinism of intelligence, it is easy to see why such an executive has been popular. The absence of some ultimate intelligent control in the brain tends to make people feel uncomfortable. As Elman and his colleagues have pointed out, "It is easy to see why such a fellow should be useful. Without him, there's a real paradox: if something is not in control of behavior, then why is behavior not uncontrolled?"

However, there is no evidence of "such a fellow" in the brain. It has been impossible to locate any area of the brain on to which all others converge as an ultimate decision center. In fact, the existence of such a center would entail another problem: if all areas report to a single, master cortical area, who or what does that single area report to? What regulates the executive?

With the waning of support for such a view, it is, perhaps, not surprising that the feature detection and convergent association theory of perception and conception, based on rigid receptive fields, is now being strongly called into question. We will return to this matter later. For the moment, though, we can see that attempts to link cerebral functions and intelligence through a mysterious "executive function," sitting on top of a hierarchy of lesser functions, will not work.

Cerebration in Context

I suggest that one reason for our failure to fully grasp what the human brain is for is the tendency to attempt to reduce intelligence to a function at a single level—for example, in a genetic program, in speed of information processing, or in the computations of reflex demons. Even Piaget and Vygotsky have been accused of one-sidedness, the former for treating the environment only as a source of disturbances to equilibria and Vygotsky for treating the brain and cognition as a "black box" into which external social relations are simply internalized.

This is not to deny that contributions to the understanding of intelligence can be made at different levels. We are certainly interested in the effects of structural genes when their contribution goes wrong, as in single-gene mutation effects. Likewise, we are interested in studies of the epigenetic interactions with cognitive regulations, such as the biochemical underpinning of learning and memory, and the bio-chemical responses of neurons to environmental change. As the British neurobiologist Steven Rose points out in his book *The Making of Memory*, we can do this without being reductionist, as the problems only start when we say intelligence resides in one or the other of these levels. Each of these studies can make sense only in the context of a general framework that can interrelate them. This is why the idea of integrated, interacting levels of regulation seems so important to the understanding of intelligence. I will now go over some implications of this view of the environment of intelligence.

Using Information

The brain is for the organization of behavior, itself a mechanism for dealing with the rapid environmental changes that can occur through-out life (not just during early development). The immediate problem for the brain is that the world is not experienced as discrete chunks of information. The information received by the sensory receptors is almost always novel, incomplete, mobile, and so on; even common features don't just spring out at us. This suggests the nature of the information used by the brain. The world can be providential for life at any level only if it contains some sort of covariation structure, through which it can be rendered predictable. However, by definition, the more complex the world into which animals have evolved, the more that correlational structure has had to be found at greater infor-mational depth. By this, I do not mean in terms of direct associations between, say, a location and food availability, but in the way this asso-ciation depends (in a complex world) on some other factor, such as time of day. I have called these deeper informational structures "hyper-structures," and argued that any system can become intelligent only by representing these in some way, and then using them as a basis for ren-dering predictability from current experience.

The real function of cognitive regulations is to cope with the way these hyperstructural relations can themselves change, sometimes rad-ically, throughout life. This may be particularly the case where animals themselves are acting on and changing their environments. Animals living in such conditions have reached a new plane of unpredictability. Any available covariation structure is not only deep, but can itself change in form throughout life. As this kind of complexity has increased, so the system for extracting that structure has needed to become more complex. The ability to do so, and to set up equally complex representational hyperstructures, is what I suggest accounts for the expansion of the brain, and particularly the cerebral cortex, in mammals. I suggest that this offers the best definition of intelligence at that level.

Finally, of course, I suggested in chapter 7 that the unique level of social cooperation emerging in human evolution resulted not only in a new order of informational depth and complexity, with which we have had to cope, but it also resulted in an entirely new kind of intelligence. Either way, if this analysis is at all correct, it tells us about the kind of information the evolved cerebral cortex should be interested in.

Hyperstructures in the Brain

Associative convergence of cerebral codes, from those of component features to increasingly complex, combined codes, such as those for objects, has long been a popular candidate for the language of the brain. In *How Brains Think*, William Calvin talks of "spatiotemporal codes" for the sight and feel of a comb, for example, which become associated (along with others) in convergence zones. But we now know that, insofar as they relate to stable patterns, such features or codes are mythical. They would need to be, simply because two views of a comb in real experience are very rarely identical. Similarly, we know that there is no single convergence zone in the brain to which all others report. Having evolved to deal with change and variability, the language of the brain has to be something other than discrete codes for features, objects, and so on.

The recent development of techniques for recording from interacting groups of neurons has led to the accumulation of evidence that what the brain is interested in is the deep covariation structures in experience, rather than this or that feature or object. British neuroscientist Donald Mackay was perhaps one of the first to emphasize the role of covariation units on the basis of neurophysiological and behavioral evidence. He argued that when the system maintains a complex representation of the world, it is registering not features, images, or other "coded" symbols, but covariations. Indeed, much of sensory behavior (such as hearing while looking, moving the eyes and head to sample a range of views, and tactile exploration while seeing) has the purpose of collecting sample covariations, he argued, rather than preformed images. He pointed out that, in the presence of analyses of covariation, even very sparse sensory input can give rise to complex form. For example, a blind man can recognize the location and shape of a manhole cover by probing it with the tip of his cane, and a coat button can be perceived in considerable detail by exploring it with the fingernail.

That this general principle seems to operate at all levels of processing from sensory input up has become increasingly clear. The Cambridge neurophysiologist Horace Barlow stressed in the 1980s that "the eye was not so much a detector of light as a detector of patterns created by those objects and events in the environment that were important for the animal," and that "exploitation of the redundancy in the input that results from the complex structure of associations it contains must play an important part in the process." We should be aware here that "redundancy" is another word often used for covariation.

This, in turn, has drawn attention to the way neurons act together at the cellular level. There is growing evidence of cells in the cortex being sensitive not only to specific inputs but in relation to what other cells are doing at the same time. A given input to a cell can have very different effects, depending on other inputs active at the same time, and the combined effect of two inputs can be much greater than their sum. Correlated activity between groups of neurons widely dispersed in the cortex are now well known, and such cooperativity has been associated with changes in what individual cells respond to (their so-called receptive fields). In other words, the activity of at least some neurons is dependent not simply on independent inputs, on an If-Then basis, but is conditioned by what other neurons are doing or have recently done.

Such findings are rapidly changing the idea of neurons (or at least some of them) as fixed feature detectors. For example, in one study, Charles Gilbert and Torsten Wiesel of Rockefeller University explored the response properties of cells in a cat's visual cortex that had previously been labeled as orientation selective, which means it has most sensitivity to lines in a specific orientation. They found that the cells were actually sensitive to interactions between disparate points in the visual field, and such interaction could actually produce alterations in the functional specificity of the cells. They pointed out how this much more powerful processing capability raises new possibilities for how the cortex analyzes visual information which "cannot be understood in terms of the properties of a single cell, but instead requires considering the properties of a neuronal ensemble."

In a more recent review, Gilbert explained how these findings require a new way of thinking of receptive fields: they are modifiable by context, experience, and expectation, rather than being fixed feature detectors. In the 1995 *Annual Review of Neuroscience*, Norman Weinberger of the University of California cites a wide range of evidence to suggest that this kind of plasticity is a lifelong capacity, and that it "constitutes a severe blow to the hypothesis that cortical perceptual functions are based on static properties of individual cells."

There is abundant evidence, in other words, that what the most evolved cerebral or cognitive systems are most interested in are covariations, simple and complex. That they are not interested in stable features or images is suggested in another, rather startling, way. As we have already seen several times, such images are extremely rare in normal experience. Even when we are standing perfectly still, for example, small but rapid oscillations of the eyeball are constantly shifting the image on the retina. When a perfectly stable visual image is presented experimentally (a technically difficult feat achieved by attaching an image to a kind of contact lens), the result is not a perfectly formed copy in the perceptual or cognitive system. Instead, the very opposite happens: the image disappears. As Mackay explained: "Stabilisation, even if it does not abolish all retinal signals, eliminates all covariation. If no correlated changes take place, there is nothing for analysers of covariation to analyse. If, then, seeing depends on the results of covariation analysis, there will be no seeing."

That the brain of a highly evolved animal, living in highly complex circumstances (in the sense described many times already), has such an

interest should not be surprising. If, as Peter Cariani of the Massachu
setts Eye and Ear Infirmary suggests, motion "appears to be essentia
for vision," that's just because of how the world, and our experience o
it, actually is. But the origin of most of that motion is quite differen
for humans compared with other species.

The Brain in Social Context

Intelligence is a property of individuals in all nonhuman vertebrates.
Cognitive regulations in the individual brain provide adaptability for
coping with rapid environmental change up to that point of evolution.
In humans, though, these cognitive regulations have become further
embedded in a new social level of regulation, which itself arose from
the need for social cooperation in highly changeable conditions. As we
saw in chapter 7, this cooperativity entailed new cognitive demands in
several ways, including attention to fine-grained detail of sound and
gesture; of contingency among these (as in language); of the physical
properties of objects (as in toolmaking); of the intentions and movement of others; and so on. Think of helping someone to move an
armoire downstairs to get some idea of the role of these factors, even
in tasks that seem mundane to a human but are well beyond the capability of any other species. Moreover, it has been part of the human
condition—partly due to the powers for making change which these
new representations provide—that these sets of relations, and those of
the physical world embedded in them, can change radically and rapidly throughout life. These demands account for the rapid and extensive expansion of the cerebral cortex in the course of human evolution.

The human brain, then, operates in what I have called a cognition-
culture complex. By evolving within a social mode of activity, the
human brain is peculiarly geared toward, and dependent on, regulations arising outside the body. These regulations vastly expand and
extend the otherwise limited powers of cognitive regulations, and thus
of the brain itself. The most recently evolved features of the human
brain, including its tripling in size, have all happened within the context of developing social cooperation. The human brain seems to have
been fashioned for that kind of lifestyle, so it functions, as Jerome

runer put it, by linking itself "with new external implementation sys-
:ms." Or, as Clifford Geertz of the University of Chicago put it even
1ore strongly:

> [The human] nervous system does not merely enable [us] to acquire
> culture, it positively demands that [we] do so if it is going to func-
> tion at all. Rather than culture acting only to supplement, develop
> and extend organically based capacities . . . it would seem to be
> ingredient to those capacities themselves. A cultureless human being
> would probably turn out to be not an intrinsically talented though
> unfulfilled ape, but a wholly mindless and consequently unworkable
> monstrosity. Like the cabbage it so much resembles, the *Homo sapi-
> ens* brain, having arisen within the framework of human culture,
> would not be viable outside of it.

As I mentioned above, a description of the part of the brain—the
cerebral cortex—that has mushroomed in the passage from apes to
humans suggests a vast replication of a basic processing unit already
present in apes. The American evolutionist Ralph Holloway has suggested that these changes in the brain could actually have been
obtained relatively easily by small changes in the epigenetic regulations
governing the rates and extent of cell division during early development. But it took the coevolution of external social structures to make
such changes viable by giving them a new purpose. However limited
the changes in one, we now know the dramatic changes they supported in the other.

Comparative psychologists have indicated how the differences in
intelligence between humans and other primates are rooted in the
presence or absence of such external implementation systems. Perhaps
even more indicative of this are the sad cases of children brought up in
various forms of isolation, either recaptured from the wild or found
locked away in attics and cellars, and thus deprived of the opportunity
to hook up to those external structures. Despite having all the genetic
and cerebral wherewithal for intelligence, they seem to exhibit its con-
spicuous absence. In his book *Seeing Voices*, Oliver Sacks describes an
eleven-year-old deaf boy who had not been exposed to, or taught, sign

language. He "seemed to be like an animal," able to make simple perceptual discriminations, but not to go beyond this. He could not hold abstract ideas in his mind, reflect, play or plan, and was "unable to juggle images or hypotheses, or possibilities, unable to enter an imaginative or figurative realm."

We must surely be grateful for the brains we have. But we must also recognize the importance of the external systems and regulations through which they obtain their proper functioning. In that interaction we develop all the cognitive functions we stress as most human: abstract knowledge and thought, language, social sensitivity, cooperativity, creativity, productivity, and so on. It has also permitted an active anticipation, and continual restructuring, of reality, so that humans adapt the world to themselves rather than adapting themselves to the world. From that interaction springs the greatest intelligence the world has ever seen.

BIBLIOGRAPHY

Calvin, W. H. 1996. *How Brains Think: Evolving Intelligences, Then and Now*. New York: Basic Books. Fairly representative account of recent "brain-centered" views of human intelligence.
Clarke, E. and K. Dewhurst. 1996 (2d ed.). *An Illustrated History of Brain Function*. San Francisco: Norman. A wonderfully illustrated account of attempts to find cognitive functions in the brain from Ancient Greece to the origins of modern scanning techniques.
Donald, M. 1991. *Origins of the Modern Mind: Three Stages of the Evolution of Culture and Cognition*. Cambridge: Harvard University Press. A detailed account of the evolution of the system which hooks up the brain to external structuring and implementation systems.
Gilbert, I. D. 1996. "Plasticity in Visual Perception and Physiology." *Current Opinions in Neurobiology* 62: 269–74. A short review of the changing perspective on neuronal specialization in cerebral cortex.
Greenfield, S. 1997. *The Human Brain: A Guided Tour*. New York: HarperCollins. A clear and candid view of knowns and unknowns about the brain.
Johnson, M. H., ed. 1993. *Brain Development and Cognition: A Reader*. Malden, Mass.: Blackwell. More recent and often interactive views on the same issues.
Mackay, D. M. 1986. "Vision—the Capture of Optical Covariation." In J. D. Pettigrew, K. J. Sanderson, and W. R. Levick, eds., *Visual Neuroscience*. New York: Cambridge University Press. A classic paper offering the radical view that the

function of the visual system is not simply a point-to-point reconstruction of the current visual scene but the abstraction of the covariation structure within it.

Rockel, A. J., R. W. Hiorns, and T. P. S. Powell. 1980. "The Basic Uniformity in Structure of the Neocortex." *Brain* 103: 321–44. The classic short statement on the lack of strong or clear correlation between structure and function in cortical tissue (expanded upon in a more contemporary statement offered by Jeffrey Elman et al. referred to in chapter 4).

Rose, S. 1993. *The Making of Memory: From Molecules to Mind.* New York: Anchor. A personal account of research on the biochemical/epigenetic bases of memory formation in the context of an antireductionist framework.

Trevarthen, C., ed. 1990. *Brain Circuits and Functions of the Mind: Essays in Honor of Roger Sperry.* New York: Cambridge University Press. A wide-ranging, and often circumspect and critical, set of articles illustrating the tortuous issues surrounding attempts to localize functions of human intelligence in the brain.

Epilogue

Promoting Human Intelligence

For thousands of years, a concept of intelligence has existed as a kind of phantom agent, "seeded" within us (by God or by genes), executed by our brains, pervading animal life, puffed up in humans, and emerging in different individuals to different degrees. We have seen that human intelligence does not exist in genes, or brains, or social environments alone, but in the complex interactions among them. In cognitive systems, generally, internal representations have the task of attuning themselves to external environmental structures that may change frequently throughout life. In humans, the evolution of a social cooperative lifestyle vastly magnified the intricacy of such structures. This required a new system of intelligence, existing in what I have called a cognition-culture complex, and needed a tripling of brain size to support it. It is a system in which social organization provides the structures to which cognitive systems (and brains) must attune, but in such a way that an individual can reflect back upon those external systems and continuously restructure them. Human intelligence resides in this dialectical relation between cognition and culture.

You may well wonder whether this view has any implications beyond that of academic debate. Does it suggest any new ways in which intelligence can be promoted? In this final offering, I suggest that it does.

"Brutal Pessimism"

The banishment of phantoms has been one of the aims of this book. Without a doubt, the most influential of these has been the g—a kind of all-round energy or power—which IQ testers say is what is being measured by their tests, is said by them to be the major determinant of success in school, as in life generally, yet remains highly obscure. Some of them, indeed, have decided that it doesn't exist as a single entity at all, but as two things, several things, or even up to eighty things in a recent count.

I was concerned in chapter 2 to describe the circular and self-fulfilling methodology that characterizes the construction of IQ tests and the widespread causal interpretations of simple correlations. Such activities are not just scientifically questionable: they have also loaned themselves to ideological propaganda and political decisions that have had deep social consequences. But behind it all has been a widespread pessimism and fatalism about the possible degrees or levels of people's intelligence. In the 1920s, Francis Binet, the author of the first IQ test, complained about the "brutal pessimism" motivating the ways in which his tests were being used. As Sandra Scarr, president of the American Psychological Society, put it in 1993: "A theory of general intelligence fits better a culture and period of relative pessimism about human perfectibility." Such views about the limits of human intelligence are regularly heard on both sides of the Atlantic.

The pessimism operates directly when decisions are made on the basis of an IQ test score that forecloses children's and other people's access to advantaged educational or occupational opportunities. But it operates more subtly, and perhaps more effectively, in numerous indirect ways. The genetic determinism at the core of IQ theory, and which forms a persuasive, if highly simplistic, picture of human intelligence, has pervaded not just the minds of psychologists but also the minds of those who operate the institutions of education and employment, as well as the general public. This idea has been hammered home throughout the twentieth century. In a recent article entitled "The g Factor and the Design of Education," the American psychologist Arthur Jensen—who has long argued that both social class and "racial" differences in IQ and school learning are partly due to genetic differences—puts it starkly. He says there are wide and "unyielding" differences in learning ability in schools; that these are largely determined by levels of g; and that differences in g are, in turn, "rooted in biology."

Not surprisingly, then, teachers in schools, educational theorists, and government ministers regularly assume a phantom biological intelligence to be operating among their pupils. They speak in terms of pupils achieving their own potentials and levels of achievement. Teachers, who have been trained in genetic IQ theory, look for signs of innate intelligence (usually, as research has shown, on the basis of social criteria, such as language, forms of self-presentation, and even facial and other physical appearance) almost from the moment pupils enter school, and soon label them as "bright" or "dull." It is now well known that such social attributions determine self-concept and self-esteem, and that these in turn largely determine children's motivations and behaviors.

The American psychologist Albert Bandura has shown how such an attribution process has spread into a wider circle of pessimism, which starts with parents' low aspirations for their children (itself socially inherited and reinforced by IQ theory), poor motivation of their children in school, low school achievement, further attributions of limited mental ability (which individuals themselves come to assimilate and accept), perception of limited opportunities, and the alienation that follows. As other research has shown, a substantial proportion of our young people—perhaps even a majority—have come to believe that they are actually incapable of learning anything very serious or complicated. I believe it involves no great stretch of logic to suggest that this cycle of attribution and pessimism accounts for no small part of society's social problems, including dropping out, a sense of hopelessness, avoidance of powerful roles in society, and the formation of dependency relationships with the institutions of the state.

This process does much to preserve the major social class structures, but also acts as a major impediment to democracy. In one survey, British psychologist Harry McGurk found this:

> The overwhelming majority of British youth appear to be politically illiterate. They have no conception of the structure of society, of how the economy works, of the characteristics of different political sys-

tems; and they are hardly aware of the policy issues, let alone the philosophical differences, which distinguish the principal British political parties.

I feel sure that future generations will come to see this as a major contemporary tragedy. Little wonder, then, that I suggest that the biggest immediate boost we can make to intelligence is to lift the yoke of pessimism which IQ theory has had placed on Western societies. IQ tests—which fail to clarify what they are measuring, add little to what teachers can already tell us about pupils in school, and have virtually no connection with complex cognitions in the outside world—should be banned. If this seems a drastic step, it is one that has been called for many times this century. In a paper in 1985, the internationally renowned psychologist Jacqueline Goodnow called for a moratorium on IQ testing, at least until we are clearer about what is being tested.

Genes and Intelligence

Much of this pessimism comes from popular beliefs about the role of genes in intelligence. Genes exist in the minds of the general public, as psychologist Susan Oyama has said, as quasi-cognitive agents, fundamentally determining our fates. This conception has no doubt been inspired by some geneticists, as well as by IQ theorists. But, as I tried to show in chapter 3, genes exist only as reactive chemicals which play a role as resources in the developmental processes from which our bodies and brains emerge. To be sure, mutations to some genes can have a devastating effect on development. But this is not the same thing as saying that they determine the form and variety of developmental end points when they are present. A wheel nut that falls off a bus can have a devastating effect on its progress, but we wouldn't thereby attribute the nut with the power to determine bus routes and variation in them when it is present.

Obviously, research aimed at identifying such defective genes, and ameliorating their consequences, is laudable work. But they fortunately affect only a very small proportion of most populations. For the rest of us, our relations with our genes is much healthier. It is, of course, an aspect of natural selection that the random genetic variation that would otherwise accumulate becomes progressively reduced and more organized. For example, one aspect of the evolution of characters important for survival is that genes associated with them form into cooperative teams. We know this does not necessitate actual reduction to genetic identity among individuals (although humans, like most species, are remarkably alike genetically, sharing more than 99 percent of their DNA), as the developmental system in the genome can use varieties of genetic material in reaching the same, or functionally equivalent, end points. Such a system is one in which development itself has started to take over from "blind" genetic selection as the method of survival in changeable environments (and thus exhibits the seeds of an intelligent system). The buffering against both genetic and environmental changes already present in the genome is vastly augmented in epigenetic, cognitive, and sociocognitive regulations. In this nested system, interaction between levels, rather than the constitution of the levels themselves, is what is important: all intelligence emerges from them.

The upshot of all this for the nature-nurture debates surrounding intelligence is quite clear in my mind. With the exception of the rare pathological conditions mentioned earlier, the idea that we could, even potentially, discover or describe a one-to-one relation between arrays of genes on the one hand and a ladder of intelligence on the other—or even that such a relation exists—seems to be highly naive. Such a relation exists it seems to me, not as a law of nature but as a law of social ideology. The fact that it is highly dysfunctional both to individuals and society, creating widespread pessimism about possibilities for intelligence, ought to give us serious pause for thought, as well as starting the search for the alternative perspectives about which this book has largely been written.

A New Intelligence

The origins of specifically human intelligence in cognition-culture interactions means that it will develop fully only when keyed into or hooked up to external cultural tools. This is what our brains and cognitive system have been prepared for in the spectacular leap in evolution from apes to humans. By cultural tools, I mean, of course, not just hardware and technological tools, but all the economic and administrative institutions through which our complex societies operate. Being fully intelligent means having access to, and cognitive grasp of, those cultural tools, and the powers through which they operate. This doesn't mean that everyone should become directors of companies and financial houses, government ministers, or directors of education or social services. But it does mean that everyone should have a clear understanding of the deliberations and operations of those institutions, and how they can have a say in them and are affected by them. And it also means that individuals are not restricted by the kind of ideologies mentioned above from fullest participation in them at all levels.

We now know much more about the fecundity of human intelligence when access to cultural tools is untrammeled by personal doubts or social ideology. A wave of observational and experimental research over the past twenty years or so has revealed what should have been clear to us already: that learning of great complexity occurs in everyday social situations, all the time, even among very young children. Examples include the dazzling acquisition of language and other social skills in the first three years of life in all children; the learning of complex technical and manufacturing skills among tribal groups; and the rapid development of literacy and numeracy among very young street traders in South American cities.

Such intelligence seems to be possible in these situations because the wider understanding of social purpose provides the scaffolding, or framework, within which current learning can be assimilated. Because all human activity is embedded in social contexts in which contingencies are highly conditional—in the sense described in chapters 6 and 7—any individual can make sense of current activities, and any learning they involve, only by being attuned to those deeper conditions. Like finding pieces for a jigsaw puzzle, you can fit them more easily and quickly when you know what the total picture is. Indeed, without such a picture, the activity becomes a nonsense.

Unfortunately, this is more or less the position in which most people are placed in their school learning and later work in modern soci-

eties. This constitutes another huge impediment to the development
of intelligence. Let me try to illustrate this by considering the two most
important institutions in this regard: education and employment.

Education

The education system is in many ways a pivot of the intellectual and
class structure of the whole of society. Year-groups of children appear
to enter the system as a more or less homogeneous beam of light into
a prism, and emerge as a spectrum of social classes reproducing that in
the society from which they entered in the first place. To most people
this seems to be a natural process. In the twentieth century, and even
as we enter into the 21st century, schooling has come to be seen by
almost everyone as the ultimate test of intelligence. To the general
public, perception of school is that of an objective and fair natural
selection process, in which children, and their innate potentials, get
sorted out by being asked to learn a neutral curriculum.

The reality, of course, is much different. First, children enter school
already advantaged or handicapped by the social inheritance of preconceptions of their own likely abilities. Then the learning is far from
neutral. The usual school curriculum, unfortunately, is the opposite of
socially meaningful contexts for learning. Children are not asked to
learn the matters governing the lives of their parents and communities:
the laws of motion of the local and national economy; the management of resources, people, and processes within them; local and
national political administration; the structure and functions of institutions; and so on. Instead, they are asked to learn "subjects," which
may seem fair enough, except that most of the subject matter that children learn in schools is not knowledge as we know it in scholarly circles, or as it is used in practical contexts. It comes in detached forms,
specially packaged to suit the administrative constraints of school and,
of course, the ideological preconceptions already mentioned.

The learning that is then used as "evidence" of children's innate
intelligence is of a very peculiar and moribund kind. Michael Cole of
the Laboratory of Comparative Human Cognition in California suggests that school learning involves, in the main, large amounts of frag-

mented information to be committed to memory; basic communication and computation (which children could probably learn more easily in wider contexts, anyway); and certain forms of knowledge classification. This problem has only become worse by recent changes. As British educationist Raymond Meighan has warned, "the 'squirrels and nuts' theory that underpins the [British] National Curriculum, whereby young people bury endless chunks of information in their memories in case it might just come in useful some day, is obsolete." Jerome Bruner has pointed out how it is the way children are subjected to artificial, made-up subjects that are not embedded in cultural practice which renders most school learning tedious and irrelevant. In other words, schools seem to be specifically set up to test children's perseverance and learning confidence, which, in turn, is a reflection of their sense of personal cognitive efficacy and their social class background.

The consequences of this pretend learning and pretend achievement for most children's intelligence are plain to see, but are usually overlooked in what seems to be a national conspiracy to do so. For example, when asked to translate their objectified knowledge back into practical contexts, even the highest school achievers seem incapable of doing so. This is seen especially in school science. In his book *The Unschooled Mind*, Harvard educationist Howard Gardner describes the results of a large number of studies on both sides of the Atlantic, as follows:

> Perhaps most stunning is the case of physics. . . . [S]tudents who receive honors grades in college-level physics are frequently unable to solve basic problems and questions encountered in a form slightly different from that on which they have been formally instructed and tested. . . . Indeed, in dozens of studies of this sort, young adults trained in science continue to exhibit the very same misconceptions and misunderstandings that one encounters in primary school children. . . . [E]ssentially the same situation has been encountered in every scholastic domain in which inquiries have been conducted.

Little wonder, then, that achievement in schools shows little relation with subsequent performances in university and/or in the real world. It has always been worrying to university admissions tutors, for example, that A-level performance seems to show little relation with how well the same students do at university. Most studies show that they account for less than 10 percent of the variability in success at university. It is perhaps not surprising, in view of this, that many studies have shown that there is little relation between academic potential and performance in the workplace at any level, even, ironically enough, for future academics! It all seems a remarkably accidental process. Even educators will sometimes admit that, after a century of compulsory education, they have little idea of what really brings about high educational achievement! Yet Cole has expressed his worries that the whole tacit model of ability on which schooling is based is now being rapidly extended across the whole world, with frightening consequences for human underfulfillment.

Many commentators have pointed out that promoting learning and intelligence in schools will require radical alteration of their objectives and the nature of curricula. British psychologist David Wood, for example, has frequently stressed how what goes on in schools needs to be more carefully aligned with what we now know to be the optimal conditions of learning, namely that which is tied in with meaningful cultural activity. This would mean, not the rote memorizing of "dead" subjects, but of how these arise in current economic, administrative, and other institutional activities. One way in which I have suggested that schools can reconnect the cognitive systems of individuals with the cultural systems in which they are immersed is through a more active system of cultural involvement. For example, local producers and practitioners can be invited to submit genuine problems to the schools, requiring thought, knowledge research, and practical solution. The news agent may have a delivery organization problem, the parish council a reporting problem, the engineer a component design problem, the steelworks a marketing problem, the health center a health-education problem, the shirt factory another kind of design problem, the farmer all kinds of botanical and zoological problems, and so on.

Within such real cultural activities, all the aims and objectives of any accepted curriculum—the development of skills of literacy and numeracy, literature and scientific research, computer use, local and national history, geography, physics, biology, design, commerce, and so on—could be worked out. But they would be worked out in meaningful contexts that would not only help develop abstract concepts in a grounded way but also engender economic sense, a sense of activities in schools being worthwhile, as well as civic identity and responsibility. It may also avoid the semi-enforced digestion of prepackaged, dead skills and knowledge, which turns school into a race of motivation and persistence, and which, in turn, stifles so much intelligence.

Employment

The cleavage of the cognition-culture relationship that occurs in schools—in which the cognitive powers of children are disengaged from the cultural tools of society, and sidetracked into rote memorization and the assimilation of fragmentary and detached pieces of knowledge—continues into employment for most adults. For most of the past two centuries, most people have simply been tied as appendages to machines and desks, their intelligence only recognized insofar as it is successfully harnessed to the production process over which few have had control, and are thus really intelligent about. This situation has been slowly changing as machines and computers have taken over the more routine tasks, and work for most people has increasingly required complex technological knowledge and wider consciousness of the general process to which they are contributing, together with the need for rapid updating of knowledge and skills. Little wonder that the British National Commission on Education (mentioned in chapter 1) can argue that "applied intelligence" has become an acute issue in national competitiveness.

As a result of these historical changes, an ironic situation has emerged. It has become increasingly realized within the business community that cultivation of this intelligence can occur only by increased "cognitive enfranchisement" of workers into the objective-setting, decision-making aspects of an enterprise. The gulf between people as labor units on the one hand and the intelligent ownership of organizations on the other is becoming widely recognized as a serious impediment to people's creativity and productivity. So moves have been made to repair the cognition-culture cleavage. As British business leader Charles Handy put it in his book *The Empty Raincoat*: "We ought instead to think of 'membership.' . . . [M]embership gives meaning and responsibility to those who work in the business. They cease to be instruments and employees and become enfranchised." Increasingly, national and international seminars are being organized by personnel officers and business leaders to find ways of incorporating workers into this repair process.

Of course, this will happen with still limited and one-sided objectives. Their significance lies in the fact that, in a period when personnel officers—with the aid of occupational psychologists (and major debates about the validity of their concepts and instruments)—have been increasingly armed with ability and aptitude tests for the management of employees, historical economic forces, of necessity, may drive out the concepts on which such tests are based. It remains to be seen what this will mean for the way intelligence is viewed by the general public; whether it will nurture a more positive conception of intelligence; and how this may feed back into what happens in schools.

Superior Intelligence

Of course, I hear you say, what about those individuals who have achieved spectacular success? The so-called gifted children or child prodigies; those who show remarkable talent in a particular domain; and even the presence among some autistic persons or other *idiot savants* (as they have been called) of stupendous calculating and artistic abilities? Doesn't their existence suggest the presence of some special genetic or environmental factors which, if we could recognize them, might help us to capitalize on them as a social and national asset? Couldn't we use tests that would identify such children at an early age, as pressure groups have suggested? Isn't it worth pursuing genetic research that could identify the crucial alleles? Can we promote experimental educational programs that could boost such abilities? My answer is that such cases certainly are intriguing, but that psychologists have little understanding of them, even in terms of how to describe the ability itself, let alone their bases.

What does seem important to me is that we tackle these issues at

the appropriate level, and view them as products of multilevel, interactive development, not single factors at single levels. Let me dispose
of two ideas right away. The idea that complex, abstract abilities, or
talent, can somehow be found in the linear code of DNA, and with it
the notion that genes act as such independent units, is as superstitious
as the idea that they can be implanted by passive experience or instruc-
tion. Such beliefs demean the system of development we have evolved.
In addition, no such tests exist which have the implied predictive ability, largely because, in my view, the nature of such abilities has been
misconceived. Instead of being mysterious agents within, awaiting the
right conditions for coming to full bloom, they are developmental
products of dialectical relations in a cognition-culture complex (as I
tried to show in chapter 7).

A number of myths pervading these questions have been duly cor-
rected by objective research. Much of the research in this area was
reviewed by Anders Ericsson and Neil Charness in a paper in the journal *American Psychologist* in 1994 (see bibliography for chapter 2).
Contrary to popular impression, they point out, few child prodigies
live up to their promise in adulthood, and most adults of exceptional
ability did not stand out as children. In addition, laboratory studies of
idiot savants (such as "calendricals," who can quickly state the day of
the week corresponding with any given date, or rapid calculators who
can quickly state the product of long multiplications) have revealed
not mysterious talents but acquired methods built on existing cultural
tools. Such performances are then reproduced in ordinary individuals
after a short period of training in the same methods. Ericsson and
Anders say, "It is curious how little empirical evidence supports the talent view of expert and exceptional performance." Instead, it is found
that such individuals have invariably been through periods of intense
efforts to learn, involving high levels of commitment and self-sacrifice
over long periods of time, usually with the support of parents who
have given up almost everything to ensure that the specific ability is
developed in their child or children. Studies tend to suggest, in other
words, just how far the human developmental system can go, given
appropriate learning conditions.

Indeed, individuals of extraordinary achievement in fields such as science are only too happy to stress the cultural and historical roots of which their work is an interactive product. Albert Einstein, for example, who did not stand out as a school pupil or university student, was only too ready to point out in his biographical notes how the previous theories of Maxwell and Lorentz "led inevitably to the theory of relativity," and insisted that the work of the individual is so bound up with that of scientific contemporaries that it appears almost as an "impersonal product of the generation."

All this suggests that no one can develop high achievement in any area without immersion in the deep cultural and historical circumstances that condition it and give the area its current form and structure. Without being able to react to that structure, individual cognitions will always be limited to the routine reproduction of limited aspects of it, rather than making original contributions that develop and change the area, as well as themselves, in the process. No doubt future research in the area will be fascinating and important. But it will not be easy. One of the major tasks will be to relinquish the simplistic preconceptions of ability that have hitherto been so sterile. The difficulty of devising tests of present or future potential, for example, is grossly underestimated because of the way they would need to describe not a fixed entity, but the complex status of a cognition-culture product which is always changing and developing. On the other hand, this is the source of the uniquely human intelligence that has so far had a spectacular history, but which faces even bigger challenges in the future.

Index